Adverse Effects and Bio-interactions of Ayurvedic Plant Drugs

A Scientific Synopsis

The Editor

C.P. KHARE (1932-)

Herbal Historian

Founder President
Society for New Age Herbals

ResearchGate

Published works cited in more than 2000 research papers.

Chandrama Khare |
Achievement
researchgate.net

Worldcat.org Overview: (RS 180.13, 615.32109540)
9 works in 51 formats in 1 language (English) and 985 library holdings (worldwide)

World Healing Plants for Tomorrow
First edition, published in 2018 in English, Astral International

Ayurveda at the Turning Point,
First edition, published in 2017 (Katiyar and Khare) in English, Kruger Brentt

Ayurvedic Pharmacopoeial Plant Drugs: Expanded Therapeutics
6 editions, published between 2015 and 2016 in English, CRC Press

The Modern Ayurveda: Milestones beyond the classical age
10 editions, published in 2012 in English, CRC Press

Dictionary of Synonyms: Indian Medicinal Plants
2 editions, published in 2012 in English, IK International

Indian Medicinal Plants: An Illustrated Dictionary
18 editions, published between 2007 and 2011 in English, Springer

Indian Herbal Remedies
9 editions, published in 2004 in English, Springer

Encyclopedia of Indian Medicinal Plants
2 editions, published between 2003 and 2004 in English, Springer

Indian Herbal Therapies: Applications of Research Findings
2 editions, published between 2000 and 2004 in English, Vishv Vijay.

Adverse Effects and Bio-interactions of Ayurvedic Plant Drugs

A Scientific Synopsis

Editor

C. P. Khare
Herbal Historian

2019

Astral International Pvt. Ltd.
New Delhi–110 002

Cover Image:
Datura stramonium Linn.
© Karen Bergeron.
Karen@altnature.com

© 2019 EDITOR
ISBN: 9789388173544 (Int. Edition)

Published by   :   **Daya Publishing House®**
                                 A Division of
                                 Astral International Pvt. Ltd.
                                 – ISO 9001:2015 Certified Company –
                                 4736/23, Ansari Road, Darya Ganj
                                 New Delhi-110 002
                                 Ph. 011-43549197, 23278134
                                 E-mail: info@astralint.com
                                 Website: www.astralint.com

Digitally Printed at   :   **Replika Press Pvt. Ltd.**

Foreword

The increase in the use of Ayurvedic/herbal medications along with modern medicines in prescribed/unprescribed form has increased the chances of both adverse drug reactions and herb-drug interactions. In spite of the known fact that there is an inadequate information and under-reporting of both adverse drug reactions and herb-drug interactions, numerous case reports have been published in recent past, which point towards the fact that herb-drug interactions certainly do occur with Ayurvedic medicines/ medicinal plants and these are much more common than we would like to believe.

In fact, the actual number of cases may be higher due to under reporting. In light of this there are many unknown plants which we are still unfamiliar with and with the list of new drugs increasing every day, the area of herb-drug interaction remains unknown. It is now known that millions of patients take herbal and conventional medicines concomitantly, often without the knowledge of their physicians. Considering our present lack of understanding of herb-drug interactions, proper reporting of such cases, careful vigilance, evidence-based appraisal and constantly updated reviews of such information are very important to promote understanding in this area. In many cases, mechanisms and causality are uncertain or unpredictable; in addition, inadequate information further makes it difficult to determine whether herb-drug interactions have occurred or not. Proper documentation is necessary with all relevant information, clear description of adverse event and careful exploration of alternative explanations along with consideration for a reasonable re-challenge whenever possible.

Ayurvedic/herbal medicines interact with synthetic drugs both in positive and negatively manner in two general ways: pharmacokinetically and pharmacodynamically. Pharmacokinetic interactions result in alterations of the modern drug's or natural medicine's absorption, distribution, metabolism or elimination. These interactions further affect drug action by either increasing their amount to have additive beneficial /toxic effect or by decreasing the amount of drug available to have no effect.

Healthcare professionals should remain vigilant for potential interactions between herbal medicines and prescribed drugs, especially when drugs with a narrow therapeutic index are used. Physicians and practitioners of both modern as well as of Traditional System of Medicines need proper training and awareness in this area. Although, one or two reports may not warrant an absolute contraindication to combinations of herbal and prescription therapies, precautions do need to be exercised by taking the medical history of patients during counseling sessions to obtain this information.

The promotion of the systematic and rational use of drugs requires the reporting of adverse events possibly caused by herbal and traditional medicines also.

The present book provides an overview of the clinical data regarding the interactions between herbal remedies and prescribed drugs, a good step in this direction.

Dr. Neeraj Tandon
Scientist-G and Head
Divisions of Publications & Information and Medicinal Plants
Indian Council of Medical Research
New Delhi.

The Red Signal

"Adverse Effects and Bio-interactions of Ayurvedic Plant Drugs" is based on the published scientific material available during last 4 decades. The project is inspired by Dr Michael T. Murray's bold theme:

What The Drug Companies Won't Tell You

And Your Doctor Doesn't Know.

Most of the bio-interactions, included in the text, are known to scientists and pharma companies, but are not known (or lesser known) to Ayurvedic *acharyas*. These will open new areas in Ayurvedic medicine for further research. But the problem is, whatever is written in older texts is the last word for Ayurvedic *acharyas*. Most of the properties (biological activities) of the herbs described in the classical texts need revalidation. But contemporary Ayurvedic *acharyas*, well-versed in Sanskrit, composed a number of *slokas* with full classical spectrum, supplemented additional properties and inserted them in *Ayurvedic Pharmacopoeia of India* and other scientific literature. This is a well-planned strategy to avoid revalidation and critical appraisal of classical texts. In some cases, abbreviated name of the contemporary composer is given, while in some only *svsa* (=self composed, by a person of unknown identity).

It is claimed, for the sake of survival, that traditionally used medicines are time-validated and safe, devoid of adverse effects. This misconception and changes in the basic profile of a number of Ayurvedic herbs after scientific research during post-classical period (as elaborated in the main text of this project) is bound to lead to a number of problems in promoting classical Ayurvedic drugs. We have failed to understand why the evidence-based herbal medicine and the medicine based on scriptures stand poles apart. It is a futile exercise to prove superiority of more than 2000 year-old systems over the emergence of research-based herbal medicine.

As Dr Nityanand has put it, maintaining the tradition should not imply that traditional medicines are to be produced in the same manner as in the 'remote past.'[1] Scientific studies on detoxification process, as advised in 2000-3000-year-old texts, do not specify the quantity of toxic principles already left in the drug. Claiming total detoxification might be misleading in some cases. For example, when toxic principles of *Croton tiglum* were extracted into hydrophilic solvents, phorbol (tumour promoting diterpene esters) only decreased from 5.12 per cent to 3.85 per cent.[2]

After the main document, a special chapter on detoxification of toxic herbs has been added. Classical procedures adopted by Ayurvedic *acharyas* need a scientific, unbiased review and total revamping, if safe and effective Ayurvedic drugs are to be manufactured on a large scale commercially. Central Drug Research Institute should work out detoxifying procedures which can be followed by herbal pharmaceutical industry. We have tried to suggest scientific detoxification process for a few toxic herbs of Ayurveda for further research.

The detoxification of toxic herbs is an important issue as it is directly related to posology.

All traditional procedures are to be revaluated due to 4 main reasons: (a) The plant which was therapeutically effective (2000-3000 years ago) may have undergone mutation and become ineffective due to ecological factors.[3] (b) A number of botanical equation of Ayurvedic plant drugs are wrong. For example, White Apaamaarga of *Ashtaangahridaya,* seventh century, is equated with *Achyranthes aspera.* It was recommended for promoting fertility, as well as bearing a male child, while, experimentally, its stem bark exhibited 100 per cent abortifacient activity.[4] (c) The disease pattern and healthcare needs of the country have changed significantly (due to total change in life style and working environment). Priorities were different when the texts of traditional systems of medicine were created.[5] (d) Herbal research has again and again emphasized that all herbal drugs must be standardized to the well-identified biologically active principle(s). In the USA, *Ashwagandha* root extract is standardized to 1.5 per cent withanolides and 1 per cent alkaloids.[6] *Boswellia serrata* gum extract is standardized to 65 per cent of boswellic acid.[6] And it is mentioned on the label. "Fingerprinted herb" is mentioned on the label of single powdered herb in the USA.[6]

In proprietary Ayurvedic medicine we find the mention of *Bhavaprakasha* or some other classical text or *Ayurvedic Pharmacopoeia of India* or some obscure book like *Vaidya Shabda Sindhu.* Dosage in some products are based on BPN. (Oh, the reference is of *Bhavaprakasha Nighantu* of 16th century for dosage in 20th century.)

It is risky to depend upon the claimed synergic action of 20, 40 or even 100 non-standardized plant drugs of different families from different locations.

Most of the material has been quoted from *WHO Monographs on Selected Medicinal Plants*, and *Reviews on Indian Medicinal Plants* series of ICMR (the only authentic source in India), and peer-reviewed research papers.

Before moving ahead, we would like to cordially acknowledge the contribution of scientists, researchers, authors, editors publishers and copyright holders, whose research papers, books, and journals have been cited in the text. The reader may go back to the original material for detailed information.

C. P. Khare

khare.herbalsociety@gmail.com

References

1. Dr Nitya Anand, *Herbal Drugs and Traditional Medicine, Perspective in the New Millennium,* Ranbaxy Science Foundation, 2006. (Symposium)

2. *The Wealth of India*, First Supplement Series, 2001, Vol. 2: 250.

3. Dr Ranjit Roy Chaudhury, WHO, SEARO, Publication No. 20: 9-14.

4. C. P. Khare, *Ayurvedic Pharmacopoeial Plant Drugs*, 2016: 18.

5. Dr B.N. Dhavan, The then Director, CDRI, Lucknow, Research and Development of Indigenous Drugs Symposium 1989: Strategies for Scientific Evaluations of Indian Medicinal Plants: 15.

6. *Natural Medicines Comprehensive Database*, Therapeutic Research Faculty, 2007: 1542, 1579, and 1716.

Contents

Adverse Effects and Bio-interactions of Ayurvedic Plant Drugs in Experimental and Clinical Trials

Abrus precatorius Linn.

Gunjaa, Jequirity

Abrus poisoning

Abrus precatorius was reported as actually one of the world's most dangerous plants, after *Ricinus communis* and *Daphne laureola*, due to the presence of abrin in the seed. If seed is swallowed or chewed, it will result in almost immediate death. Abrin is one of the most lethal poison, inducing severe vomiting, high fever, drooling, highly elevated levels of nervous tension, liver failure, bladder failure, bleeding from the eyes, and convulsive seizures. Abrin is a toxalbumin very similar to ricin found in castor seeds.

Toxicological analysis was reported not of much help in cases of Abrus poisoning while thin layer chromatography using the seed extract and patient's serum could be helpful. The decontamination (by stomach wash) was suggested as the major mode of treatment of abrus poisoning as no antidote is available yet. However, as the cause of death in most reported cases appears to be renal failure, hemodialysis is indicated in severe poisoning with associated renal compromise. (Victor Kuete, in *Toxicological Survey of African Medicinal Plants*, 2014.)

Abrin is a single glycoprotein of molecular weight 60,000 to 65,000; a type 2 ribosome inactivating protein. The toxin is composed of 2 chains (A and B.) The A chain (effectomer) is responsible for the toxic activity, the B chain (haptomer) binds to glucose units of cell surface carbohydrates. The A chain is responsible for the toxic activity. Once inside the cell, the A chain migrates to the 60S unit of the ribosome, acting to inhibit further protein synthesis. Abrin has a strong inhibitory effect on protein synthesis, moderate inhibitory effect on DNA synthesis, and little effect on RNA synthesis. Abrin is present to the extent of 0.15 per cent in the seed. 5 mg of arbin is reported to be toxic to humans. In goats, ground seeds 1-2 g/kg/day caused death in 2 to 5 days. The LD_{50} of abrin given to mice is 0.04 mcg.

Biological Activity of Toxic Constituents

Arbridine, isolated from the seed, has been identified as the anti-fertility principle. It affected the fertility of female albino rats. Induced100 per cent sterility in rats when injected 1 day pre and post coitum. (Zia-ul-Haque *et al.*, *Pak J Zool*, 1983, 15: 141.) The methanolic extract of seeds inhibited the motility of human spermatozoa.

The leaves contain neuronal poisonous components; ethanolic extract of the leaves possesses a similarity to *d*-tubocurarine in respect of the pattern of neuromuscular blockade. (*The Wealth of India, First and second Supplement Series*, 2000 and 2006: 6-7; 4-5.)

Abrin has been used with some clinical success as an analgesic in terminally ill patients.

In an experimental study, abrin exhibited anti-tumour activity in mice when used at a sublethal dose of 7.5 mcg/kg every alternate day for 10 days. Both intralesional and intraperiloneal administration of abrin was effective in reducing solid tumour mass development induced by Dalton's Lymphoma Ascites (DLA) and Ehrlich's Ascites Carcinoma (EAC) cells. DLA cell line was more sensitive to abrin than EAC. Abrin when injected i.p. increased the life span of ascites tumour bearing mice. (Ramnath V *et al.*, Antitumour effect of abrin on transplanted tumours in mice, *Indian J Physiol Pharmacol*, 2002 Jan, 46(1): 69-77.)

Aconitum ferox Wall. ex Serr.

Vatsanaabha, Indian Aconite

Indian Aconite is a mixture of *A. ferox*, *A. napellus* and *A. bisma* syn. *A. palmatum*

Analgesic and Antipyretic Activity of *Shodhit* Aconite

The effect of crude aconite and *shodhit* aconite (prepared by cold treatment with cow's urine and hot treatment with cow's milk) were tested for CNS activity by studying the analgesic action in albino rats and antipyretic action in experimentally induced pyrexia in rats. There was significant analgesic response in all the three samples (crude, cow's urine-treated and hot cow's milk-treated.) The antipyretic effect of all the three samples was found significant. The antipyretic action of paracetamol and the milk-treated aconite was of short duration, while the effect of the urine-treated aconite was more sustained.

Cardiotoxic Activity

The cardiotoxic effect of aconite, accelerating the heart and resulting in irregularity of cardiac activity was observed. The ED_{50} for the cardiac toxicity for milk-treated aconite was 75 mg and for urine-treated aconite was 32.5 mg, while for crude drug it was 19.5 mg. The crude drug of all the three species of Indian aconite produced convulsions, irregular breathing and other symptoms of aconite toxicity. (Singh LB *et al.*, *Indian J Pharmacol*, 1981, 13: 123-124. *Cf. Reviews on Indian Medicinal Plants*, ICMR, 2004, Vol. 1: 185-186.)

Pro-arrhythmic and Anti-hypertensive Activity

For the study, *Aconitum napellus* root (tuberous aconite) was purified by cold treatment with cow's urine. In order to claim the detoxification effect on aconite, the aconite roots was studied for its pro-arrhythmic and anti-hypertensive activity in both, raw and treated aconite. Adult wistar rats of either sex weighing between 250-300 g were randomly assigned to four groups with six animals in each group, *viz.*, normal control (olive oil 1ml, p.o.); diseased control; unpurified aconite root 50 mg/kg, p.o. and purified aconite root 50 mg/kg, p.o.

Aconite root treatment in both forms (purified and unpurified) caused significant reduction in blood pressure when compared with diseased control group (P<0.05.) The unpurified aconite root group showed significant increase in heart rate, increase in QRS complex time and increase in QT interval, however these parameters were statistically insignificant in purified aconite root treated group. This indicated that purified aconite root is devoid of the pro-arrhythmic activity. The probable mechanism of anti-hypertensive activity of aconite root can be attributed to decrease in plasma renin activity, decrease in oxidative stress and increase in nitric oxide levels. (Arindam Paul *et al.*, Effect of *shodhan* on dried tuberous Aconitum root, *Indonesian J. Pharm*, 2012, Vol. 24(1): 40-46.)

Acorus calamus Linn.

Vacha, Sweet Flag

Calamus products are among regulated products in the USA as the volatile oil contains isoasarone, the carcinogenic principle.

Genotoxicity

An Indian study assessed the sub-acute toxicity and genotoxicity potentials of *A. calamus* rhizome extract in Swiss albino mice. The mice were treated with a low (400 mg/kg) and a high (800 mg/kg) dose of *A. calamus* rhizome extract for 14 days. The sub-acute toxicity study analysed the haematological, biochemical parameters, and histopathology of liver, kidney and spleen of animals. The genotoxicity assessment was done *in vivo* by chromosomal aberrations assay on bone marrow cells of mice. No mortality or behavioural adverse effects were observed in animals by 800 mg/kg oral dose of extract for 14 days. In biochemical assay, only 800 mg/kg dose revealed significant increase in alanine aminotransferase (ALT), aspartate aminotransferase (AST), and total and direct bilirubin. No major effects were evident in any biochemical parameters by 400 mg/kg dose. In the haematological assay, only the high dose of extract showed significant increase in the numbers of neutrophils and lymphocytes, however, all blood parameters appeared normal in mice treated with 400 mg/kg dose of extract. The histopathological assessments of liver, kidney and spleen tissues also showed several deviations from their normal architectures only in the high dose treated group of animals. The genotoxic assessments revealed maximum frequencies of chromosomal aberrations in 800 mg/kg extract-treated animals. Taken together, this study suggested that 800 mg/kg dose of *A. calamus* rhizome extract may cause several toxic and genotoxic effects in experimental animals. (Arun K. Yadav, Sub-acute toxicity and genotoxicity assessment of the rhizome extract of *Acorus calamus*, *European Journal of Pharmaceutical Research*, 4(8): 392-399.)

Chemo-types of *Acorus calamus*

In literature, *Acorus calamus* has been classified into four chemo-types which are found in different locations worldwide. Type I: *Acorus calamus* L. var. *americanus*, a diploid American variety; Type II: variety *vulgaris* L. (var. *calamus*), a European triploid; Type III and IV: variety *augustatus* Bess. and Dutch variety *versus* L., subtropical tetraploids. (Chief constituents of volatile oil are heavily dependent upon the chemical strain (di-, tri-, tetraploid.) North American chemo-type I is virtually ioasarone-free. Western European chemo-type II contains less, 10 per cent isoasarone. The two other chemo-types (III and IV) have been found to contain up to 96 per cent isoasarone in volatile oil.

All the 27 Indian genotypes of *A. calamus* were also analyzed for α and β-asarone contents, and percentage of essential oil. The genotype (Ac13) from Kullu (Himachal Pradesh) showed maximum (9.5 per cent) percentage of oil, whereas corresponding minimum (2.8 per cent) was obtained from the genotypes from Pangthang (Sikkim.) Similarly, the highest α and β-asarone contents (16.82 per cent and 92.12 per cent) were obtained from genotypes from Renuka (Himachal Pradesh) and Udhampur (Jammu and Kashmir), while lowest α and β-asarone contents (0.83 per cent and 65.96 per cent) resulted from Auranwa (Uttar Pradesh) and Pangthang (Sikkim) genotypes, respectively. (T.S. Rana *et al., Physiol Mol Biol Plants,* 2013 Apr, 19(2): 231–237.)

Aegle marmelos (L.) Correa ex Roxb.

Bilva, Bengal Quince

Effect on Spermatogenesis and Fertility

In a study at the Department of Zoology and Botany, Banaras Hindu University, Varanasi, India, the effect of aqueous extracts of *A. marmelos* dried leaves (AEAM) on cyclophosphamide (CPA)-induced changes in sperm characteristics and testicular oxidative damage in mice was investigated. Thirty six adult Parke's strain mice were divided into six groups: Group I: Control mice received distilled water (intraperitoneally) once in a week for 5 weeks. Group II: AEAM (600 mg/kg bw, orally) once in a week for 5 weeks. Group III: CPA 200 mg/kg bw, for 5 weeks (once in a week) by intraperitoneal injection (The dose of cyclophosphamide was selected based on earlier studies as 200 mg/kg bw. Group IV: CPA (200 mg/kg bw, intraperitoneally) and AEAM (400 mg/kg bw) orally for 5 weeks (once in a week.) Group V: CPA (200 mg/kg bw, intraperitoneally) and AEAM (500 mg/kg bw) orally for 5 weeks (once in a week.) Group VI: CPA (200 mg/kg bw, intraperitoneally) and AEAM (600 mg/kg bw) orally for 5 weeks (once in a week.)

CPA was found to reduce gonadosomatic index (GSI), sperm counts, motility, viability, antioxidant activities and induced histopathological changes of testis. In the group administered AEAM with CPA an exacerbation of sperm count, motility and viability of the cauda epididymis, GSI, antioxidant activities and architecture of testis was observed. The results showed that the administration of AEAM may aggravate CPA-induced reproductive toxicity. (Sangita Singh *et al.*, *Braz J Pharm Sci*, 53(3): Epub Oct 26, 2017.)

Ethanolic extract of leaf shows dose-dependent decrease of testosterone levels, spermatogenesis and fertility in rats. (Chauhan A *et al.*, Suppression of fertility in male albino rats following the administration of 50 per cent ethanolic extract of *Aegle marmelos*, *Contraception*, 2007, 76: 474-481. Chauhan A *et al.*, Reversible changes in the anti-fertility induced by *Aegle marmelos* in male albino rats, *Syst Biol Reprod Med*, 2008, 54: 240-246.)

Effect on Thyroid Hormones

Relative importance of *Bacopa monnieri* (200 mg/kg), *Aegle marmelos* (1.00 g/kg) and *Aloe vera* (125 mg/kg) leaf extracts in the regulation of thyroid hormone concentrations in male mice was investigated. While serum levels of both T_3 and T_4 were inhibited by *A. vera*, *A. marmelos* extract could decrease only T_3 concentration. On the other hand, T_4 concentration was increased by *B. monnieri* extract suggesting its thyroid-stimulating role. When the relative potency of each plant extract was calculated in terms of percent increase or decrease in thyroid hormones, as compared to the control value, the decrease in T_3 concentration by *A. marmelos* was about 62 per cent indicating its possible use in the regulation of hyperthyroidism. *B. monnieri* could increase T_4 concentration by 41 per cent without enhancing hepatic lipid peroxidation (LPO) suggesting that it can be used as a thyroid-stimulating drug. In fact, hepatic LPO was decreased and superoxide dismutase (SOD) and catalase (CAT) activities were increased by *B. monnieri* and *A. marmelos* leaf extracts showing their antiperoxidative role. It is thus suggested that *A. marmelos* and *A. vera* may be used in the regulation of hyperthyroidism, while *B. monnieri* in hypothyroidism. (Kar A, Panda S, Bharti S, *J Ethanopharmacol*, 01Jul, 2002, 81(2): 281-285. PMID 12065164.)

Alangium salvifolium (Linn. f.) Wang.

Ankola, Sage-leaved Alangium

During classical period of Ayurveda, the root bark of Ankola was prescribed internally as well as externally in cases of poisonous animal bites, snake bites, dog bites, rabies, spider poison. (*Sushrut Samhita, Ashtangahridya, Chakradatta, Sharangadhara Samhita* and *Bhavaprakasha*; from 1000 BC to 16th century AD.)

Ayurvedic Pharmacopoeia of India, Part I, Vol. V, validated Ankola's use as a fish poison, and in poisonous animal bites.

Toxicity of Alkaloidal Fraction in an Experimental Trial

The total alkaloidal fraction of the leaves was administered orally to albino rats to study its anti-inflammatory effect, using formalin-induced arthritis and granuloma pouch techniques; beta-methasone was used for comparison. Though anti-inflammtory reaction was observed during 5 to 11 days, the alkaloid fraction showed toxicity. 66.7 per cent of rats died during treatment. (Prasad. D N *et al., Indian J Med Res*, 1966, 54: 282.)

Possible Role of Chemical Constituents in Poisoning

Chemical constituents of the root bark include alangin A and alangin B, alanginine, emetin, cephaeline, and betasterol. These may act as an antidote to poisoning. The menthol extract of flowers contain steroids and flavonoids which may exert antibacterial activity against both Gram positive and Gram negative bacteria. Seed alkaloids include N-methylcephaeline, emetin, cepheline and psychotrine, which exhibit antitoxic property in accidental poisoning. Emetine causes vomiting. Psychotrine shows antiplasmodial activity. Stigma of the plant contains myristic acid, E-cisfused neohophane derivatives and its isomer, N-benzoyl-L-Ph-alaninol. (Sanjeev Kumar and Chandra Shekhar, *Int J Res Ayurveda Pharm*, 8(4), 2017. *www.ijrap.net*)

Invalidated Claims of Herbal Antidotes for Poisoning

Central Research Institute of Siddha, Chennai, Tamil Nadu, reiterated that no claims regarding effect of a number of herbal drugs on snake venom poisoning could be validated either clinically or experimentally.

Siddha drugs, advocated in Tamil Nadu, were subjected to limited pilot study to prove their efficacy in cobra envenomation in albino rats. Each Siddha drug administered orally in 4 dose levels against LD_{50} of venom (0.82 mg/kg, *s.c.*) could not afford any protection against cobra envenomation, resulting 100 per cent mortality in rats.

The study invalidated the claims of herbal 'antidotes' for cobra venom poisoning. (*Pharmacological Investigations of Certain Medicinal Plants and Compound Formulations used in Ayurveda and Siddha*, Central Council for Research in Ayurveda and Siddha, 1996: 427.)

Albizia lebbeck (Linn.) Benth.

Shirish, East Indian Walnut

Effect on Spermatogenesis in Male Rats

Methanolic extract of *Albizia lebbeck* bark when administered orally at the dose level of 100 mg/rat/day to male rats of proven fertility for 60 days did not cause any significant loss in their body weights but the weights of reproductive organs, *i.e.* testis, epididymides, seminal vesicle and ventral prostate were decreased in a significant manner when compared to controls. Sperm motility as well as sperm density were reduced significantly which resulted in reduction of male fertility by 100 per cent. Marked decline in the germ cell population was noticed. Population of preleptotene, pachytene, secondary spermatocytes and step-19 spermatid were declined by 60.86 per cent, 65.81 per cent, 71.56 per cent and 66.55 per cent, respectively. Cross-sectional surface area of sertoli cells as well as the cells counts were found to be depleted significantly. Leydig cells nuclear area and number of mature Leydig cells were decreased by 60.03 per cent and 51.56 per cent, respectively. Serum testosterone levels showed significant reduction after A. lebbeck extract feeding. Oral administration of the extract did not affect red blood cell (RBC) and white blood cell (WBC) count, haemoglobin, haematocrit and glucose in the blood and cholesterol, protein, triglyceride and phospholipid in the serum. In conclusion, *A. lebbeck* bark extract administration arrests spermatogenesis in male rats without noticeable side effects. (Gupta R S *et al., International Journal of Phytotherapy and Phytopharmacology,* 19 Sept 2005, 13(4): 277-283.)

Allium sativum Linn.

Lashuna, Garlic

Safety Concern of Chopped Garlic in Room Temperature

According to Food and Drug administration, chopped garlic and oil mixes left at room temperature have the ability to result in fatal botulism food poisoning. Such products are to be kept refrigerated, especially those that do not contain acidifying agents, such as phosphoric or citric acid. Atleast 40 cases of poisoning were reported in the late 1980. (www.fda.gov. March 1989.)

Coagulation Dysfunctions

Coagulation dysfunctions have been reported, such as postoperative bleeding and prolonged clotting time. One case of spinal epidural hematoma associated with excessive garlic ingestion has been reported. An 87-year old man who reported to consume an average of 4 cloves of garlic per day to prevent heart disease, presented to the emergency room with acute onset abdominal discomfort and bilateral sensory and motor paralysis in the lower extremities. Prothrombin time was 12.7 seconds and partial thromboplastin time was 22.3 seconds. The patient had no risk factors for bleeding and was not taking any other medications that would affect bleeding tendency. The platelet inhibition caused by garlic was determined to be the cause. (Rose K D *et al., Neurosurgery*, 1990, 26(5): 880-882.)

Effect on Gastrointestinal Mucosa

In 2001, in a study, it was investigated whether different garlic preparations have undesirable effects on gastrointestinal mucosa in dogs. When administered directly to the stomach, Aged Garlic extract did not produce any changes compared to control to mucosa, whereas boiled garlic powder caused redness and raw garlic powder also caused redness and erosion of the mucosa. Pulverized enteric-coated tablets caused redness and loss of epithelial cells. Further studies should be done in humans to confirm the relevance of these findings. (Hoshino T, 131: *et al., J Nutr*, 2001: 1109S-1113S.)

"Garlic Burns" after Topical Application

A 38-year old woman developed a garlic burn from applying a poultice made from fresh, uncooked garlic to her breast for treatment of a self-diagnosed *Candida* infection secondary to breastfeeding for her 6-month-old son. Despite a burning sensation upon application, she left the poultice in place for 2 days. The infant showed no apparent adverse effects. She presented to the emergency room after removal of the poultice. The area appeared as a burn with skin loss, ulceration, crusting, hyperpigmentation, granulation of tissue, serous discharge, minor bleeding, and erythema on the periphery. (Roberge R J *et al., Am J Emerg Med*, 1997, 15: 548-549.)

Interaction with Drugs

Garlic had proved to change pharmacokinetic variables of paracetamol, decreased blood concentrations of warfarin and produced hypoglycaemia when taken with chlorpropamide (Izzo A A and Ernst., E, Interactions between herbal medicines and prescribed drugs: a systematic review, *Drugs*, 2001, 61(15): 2163-2175.)

Allicin content of the garlic can induce the cytochrome P450 3A4 (CYP 3A4) isoenzyme and can result in clinically important decreases in concentrations of drug metabolized by the enzyme. This interaction was proven with saquinavir (antiretroviral drug.) However, a garlic preparation containing alliin and alliinase (which formed one-half the amount of alliicin stated on the label) did not significantly inhibit CYP 3A4, which was proven by a lack of interaction with the drug alprazolam. (*Herbal Toxicology*, Amit Sarkar: 119.)

Aloe vera Tourn. ex Linn.
Ghritkumari, Aloe

Complications Due to Chronic Abuse

Leaf juice of *Aloe vera* must be used only for a short term for constipation. Long-term use is associated with laxative dependence, albuminurea, electrolyte imbalance, hematuria, hypokalemia, malabsorption, melanosis coli, metabolic acidosis, orthostatic hypotension, osteomalacia, protein-losing gastroenteropathy, renal tubular damage, steatorrhea, weakness, weight loss.

Existing hypokalemia resulting from long-term abuse can potentiate the effects of cardiac glycosides (digitalis, strophanthus) and anti-arrhythmic drugs such as quinidine. (*WHO monographs on selected medicinal plants*, Vol. 1: 39.)

While chronic abuse of anthranoid-containing laxatives was hypothesized to play a role in colorectal cancer, no casual relationship between anthranoid laxative abuse and colorectal cancer has been demonstrated.

Tissue Damage

Chronic treatment with high dose of Aloe reduces vasoactive intestinal peptide and somatostatin levels, which may damage enteric nervous tissue. (*PDR for Herbal Medicines*, 2007: 23.)

Hepatic and Renal Toxicity of Cape Aloes

A 47-year-old man developed acute oliguric renal failure and liver dysfunction after ingestion of Cape Aloes, a previously described nephrotoxin. (Luyckx VA *et al.*, Herbal remedy—associated acute renal failure secondary to Cape aloes. *Am J Kidney Dis.* 2002, 39: E13.)

In recent years several cases of aloe-induced hepatotoxicity were reported. The clinical manifestation, laboratory findings and histological findings of three persons admitted to the hospital for acute hepatitis taking Aloe preparation for months, met diagnostic criteria of toxic hepatitis. Upon discontinuation of the oral Aloe preparations, liver enzymes returned to normal level. (Yang H N *et al.*, Aloe- *induced toxic hepatitis*, *J Korean Med Sci*, 2010, 25(3): 492-495.)

The first case of acute hepatitis due to the ingestion of *Aloe vera* compound was reported in 2005 in Germany. Afterwards, cases of Aloe-induced toxic hepatitis were reported in Turkey, United States, Argentina, and Korea. A total of six females and two males were admitted to hospital for acute hepatitis after taking Aloe preparation over 3–260 weeks. Their clinical manifestation, liver biopsy, and laboratory findings supported the diagnosis of toxic hepatitis. All eight patients showed improved conditions after discontinuing *Aloe vera* preparations. *(Aloe vera*: A review of toxicity and adverse clinical effects, Xiaoqing Guo and Nan Mei, Division of Genetic and Molecular Toxicology, National Center for Toxicological Research, Jefferson, Arkansas, USA, *Journal of Environmental Science and Health*, Part C, March 2016.)

Effect on Thyroid Hormones

Relative importance of *Bacopa monnieri* (200 mg/kg), *Aegle marmelos* (1.00 g/kg) and *Aloe vera* (125 mg/kg) leaf extracts in the regulation of thyroid hormone concentrations in male mice was investigated. While serum levels of both T_3 and T_4 were inhibited by *A. vera*, *A. marmelos* extract could decrease only T_3 concentration. On the other hand, T_4 concentration was increased by *B. monnieri* extract suggesting its thyroid-stimulating role. When the relative potency of each plant extract was calculated in terms of percent increase or decrease in thyroid hormones, as compared to the control value, the decrease in T_3 concentration by *A. marmelos* was about 62 per cent indicating its possible use in the regulation of hyperthyroidism. *B. monnieri* could increase T_4 concentration by 41 per cent without enhancing hepatic lipid peroxidation (LPO) suggesting that it can be used as a thyroid-stimulating drug. In fact, hepatic LPO was decreased and superoxide dismutase (SOD) and catalase (CAT) activities were increased by *B. monnieri* and *A. marmelos* leaf extracts showing their antiperoxidative role. It is thus suggested that *A. marmelos* and *A. vera* may be used in the regulation of hyperthyroidism, while *B. monnieri* in hypothyroidism. (Kar A, Panda S, Bharti S, *J Ethanopharmacol*, 01Jul, 2002, 81(2): 281-285. PMID 12065164.)

Carcinogenic Activity

Recently, *Aloe vera* whole leaf extract showed carcinogenic activity in rats, and was classified by the International Agency for Research on Cancer as a possible human carcinogen (Group 2B.) This review presented updated information on the toxicological effects, including the cytotoxicity, genotoxicity, carcinogenicity, and adverse clinical effects, of *Aloe vera* whole leaf extract, gel, and latex. (Xiaoqing Guo and Nan Mei, Aloe Vera–A Review of Toxicity and Adverse Clinical Effects, *Journal of Environmental Science and Health*, Part C, March 2016, 34(2): DOI 10.1080/10590501.2016.)

Contraindications

Internally, Aloe is contraindicated in abdominal pain of unknown origin, Corhn's disease, inflamed intestinal diseases, ulcerative colitis, appendicitis, diverticulitis, gastrointestinal obstruction, haemorrhoids, hypokalemia, irritable bowel syndrome, kidney disease, and severe dehydration.

On May 9, 2002, the U.S. Food and Drug Administration issued a final rule banning the use of aloe and cascara sagrada as laxative ingredients in over-the-counter drug products. Rectal bleeding or failure to have a bowel movement within 24 hours after use of a laxative may indicate a serious condition. Chronic use may cause dependence and need for increased dosages, disturbances of water and electrolyte balance (hypokalemia), and an atomic colon with impaired function.

Although, no teratogenic or fetotoxic effects were seen in rats after treatment with Aloe extract (up to 1000 mg/kg) or aloin (up to 200 mg/kg), anthranoid metabolities appear in breast milk and Aloe is contraindicated during pregnancy and lactation.) (*WHO monographs on selected medicinal plants*, Vol. 1; 39.)

Aloe latex contains anthraquinones that may stimulate uterine muscle activity, initiate premature labour, or possibly cause abortion when given orally. (*PDR for Herbal Medicines*, 2007: 23.)

Interactions with Sevoflurane

A 35-year-old woman experienced massive intraoperative bleeding after oral consumption of *Aloe vera* tablets for two weeks before the surgery for leg pain. Compounds contained within *Aloe vera* can reduce the synthesis of prostaglandin, thus inhibiting secondary aggregation of platelets. Sevoflurane, a general anesthetic, inhibits thromboxane A(2) formation by suppressing cyclooxygenase activity. Since both sevoflurane and *Aloe vera* have antiplatelet effects, the bleeding could have been due to a possible herb-drug interaction between *Aloe vera* and sevoflurane. (Lee A *et al.*, Possible interaction between sevoflurane and *Aloe vera*, *Ann Pharmacother*. 2004, 38: 1651-1654.)

The juice of the leaves of certain species of Aloe is poisonous, for example *Aloe venenosa* is poisonous.

Alpinia galanga Willd.

Kulanjana, Greater Galangal

Effect on HIV-1 Replication

In a study, it was shown that 12S-12-acetoxychavicol acetate (ACA), a small molecular compound isolated from the rhizomes of *Alpinia galanga*, inhibited Rev transport at a low concentration by binding to chromosomal region maintenance 1 and accumulating full-length HIV-1 RNA in the nucleus, resulting in a block in HIV-1 replication in peripheral blood mononuclear cells. Additionally, ACA and didanosine acted synergistically to inhibit HIV-1 replication. Thus, ACA may represent a novel treatment for HIV-1 infection, especially in combination with other anti-HIV drugs. (Ying Ye, Baon Li, *Journal of General Virology*, 87: 2047-2053, online 01 July 2006.)

Effect on Estrogen

A study provided initial evidence towards establishing either estrogen agonistic or antagonistic hormonal property of methanolic extract of shade dried rhizome of *Alpinia galanga* (MAGE) in *in vivo* animal model. Working doses of the MAGE were determined after LD_{50} study selecting two doses of 200 mg/kg bw/day and 500 mg/kg bw/day. After 7 days of oral treatment, animals were sacrificed. Morphological observations revealed that low dose MAGE treated group showed weak estrogenic activity. The high dose MAGE treated group showed no estrogenic activity rather showed decrease uterine wet weight as well as morphologically constricted uterine horns which clearly suggests anti-estrogenic activity. (Yungkham Rajeevkumar Singh, Jogen Kalita, Effects of Methanolic extract of *Alpinia galanga* from Manipur (India) on uterus of ovariectomised C3H albino mice, *International Research Journal of Pharmacy*, May 2012, 3(5): 423-427.)

Effect on Testosterone

In a study, male rats were divided at random into 3 groups of 10 animals each. The dry extract 2.4 g and 4.8 g were dissolved separately in 100 mL of 5 per cent ethanol with resultant solutions of 100 mg/mL (solution A) and 300 mg/mL (solution B.) Group 1 (Controls) injected 2 mL/day of 5 per cent ethanol. Group 2 (low Alpinia treated) received solution A (100 mg/day.) Group 3 (high Alpinia treated) received solution B (300 mg/day.) 56 days after induction, the rats were sacrificed. Methanol extract of *A. galanga* increased serum testosterones level significantly in both treated groups in comparison with control group ($p < 0.05$.) Besides, the percentage of sperm viability and motility in both tested groups were significantly increased. Follicle stimulating hormone FSH hormone, morphology and weight were affected in both treated groups. With 300 mg/day an increase in sperm count was observed. Sperm motility was increased in two treated groups whereas testis weight was decreased in treated groups. Real time analysis of treated cells of testis showed increase level of mRNA related to CREM gene involved in spermatogenesis process after 56 days induction. (Mazaheri M *et al., Iran J Reprod Med*, 2014 Nov, 12(11): 765-770.)

Effect on Neurodegenerative Disorders

In vitro, Alpinia galanga extract inhibited the activation of THP-1 cells and reduced inflammation-related gene expression, which the investigators stated may be useful in delaying the onset and the progression of neurodegenerative disorders such as Alzheimer's disease. According to a review, some formulations that contain *Alpinia*, have been shown to modulate events or signaling pathways related to Parkinson's disease pathogenesis. (*The Journal of Alternative and Complementary Medicine*, Vol. 10, No. 6.)

Ananas comosus (Linn.) Merrill

Bahunetra (non-classical), Pineapple

Bromelain is a general name for proteolytic enzymes obtained from the stem and fruit of pineapple. It is used for acute postoperative and post-traumatic swelling, also used orally as an anti-inflammatory agent for allergic rhinitis, pulmonary edema, acute sinusitis, and smooth muscle relaxation.

Abortifacient Activity of Unripe Fruit's Steroids

Steroids isolated from unripe fruits and juice are found to exert abortifacient effect in mice. The steroid 50-stigmastane-3/3,5.6/3-triol 3-mono benzoate exerted maximum interceptive effect when administered during 6-7 days of gestation in mice. It has been suggested that the compound acts by exerting anti-lutetrophic activity. It also revealed anti-ovulatory activity in rabbits. (Pakrashi A. and Chakrabarty, 1979, cited in *Ancient Science of Life*, 1984, Vol III(4): 193-202.)

Interaction of Bromelain with Antibiotics, and Antiplatelet Drugs

Bromelain increases blood and urine levels of certain antibiotics in humans by increasing the volume of distribution of tetracycline and amoxicillin. The mechanism may be associated with enhanced absorption and increased permeability into diseased tissue, thus enhancing antibiotic access to the infected site. (*The Review of Natural Products*, Wolters Kluwer, 7th Edn. 2012: 1275.)

Conraindications

Bromelain is contraindicated in patients who have severe liver or kidney impairment or who need dialysis.

The supplement should avoided by patients who have a coagulation disorder such as haemophilia.

Bromelain is capable of inducing IgE-mediated respiratory and gastrointestinal allergic reactions. Human subjects can exhibit a cross-reaction between the two plant proteases bromelain and papain (an enzyme of *Carica papaya*.)

Bromelain may increase heart rate at higher doses. It should be used cautiously (doses less than 500 mg per day) in patients with heart palpitation or tachycardia. (*PDR for Herbal Medicines*, 2007: 947.)

(Commercial Bromelain is not a chemically homogeneous substance because if the enzyme is highly purified it loses its stability and most of its physiological activity. The proteolytic enzyme also contains small amounts of an acid phosphatase, a peroxidase, several protease inhibitors and organically bound calcium.)

Andrographis paniculata Wall. ex Nees

Kaalmegh, Creat

Adverse Effect on Spermatogenesis

One study by Akbarsha M A, Manivannan B, Shahul Hamid K, indicated that andrographis may impair fertility. Dry leaf powder of *A. paniculata*, when fed orally to male albino rats, at a dose level of 20 mg powder per day for 60 days, resulted in cessation of spermatogenesis, degenerative changes in the seminiferous tubules, regression of Leydig cells and regressive and/or degenerative changes in the epididymis, seminal vesicle, ventral prostate and coagulating gland. There was reduction in the weight and fluid content of the accessory glands. The treatment also resulted in accumulation of glycogen and cholesterol in the testis, and increased activities of lactate dehydrogenase in testis and alkaline phosphatase in testis and ventral prostate. The results suggest antispermatogenic and/or antiandrogenic effect of the plant. (Akbarsha MA *et al.*, Antifertility effect of *Andrographis paniculata* (Nees) in male albino rat, *Indian J Exp Biol*,1990 May,28(5): 421-426.)

Spermotoxic Effect of Andrographolide

A follow-up of the earlier, 1990, study was undertaken to investigate whether andrographolide, one of the major constituents of the plant, is responsible for such an effect. The compound was administered to 3-month-old male Wistar albino rats at two dose levels, for 48 days. Fertility tests, analysis of the counts, motility and abnormalities of the cauda epididymidal spermatozoa, and histopathological-evaluation of the testis were carried out. The results showed that sperm counts decreased, the spermatozoa were not motile, and several of them possessed abnormalities. The seminiferous epithelium was-thoroughly disrupted and in the seminiferous tubules, fully differentiated spermatozoa were far too limited; cells in the divisional stages were prevalent; multinucleate giant cells were abundant and Leydig cells appeared intact. It is inferred that andrographolide could affect spermatogenesis by preventing cytokinesis of the dividing spermatogenic cell lines. The multinucleate giant cells are comparable to the symplasts generated by cytochalasin-D and ursolic acid due to action at stages V-VII of the spermatogenic cycle. Sertoli cell damage and spermatotoxic effects are also apparent. Thus, the study points to a male reproductive toxic effect of this compound when used as a therapeutic; the study also confirms the possible prospective use of andrographolide in male contraception. (Akbarsha MA, Murugaian, Aspects of the male reproductive toxicity/male antifertility property of andrographolide in albino rats: effect on the testis and the cauda epididymidal spermatozoa, *Phytother Res*, 2000 Sept, 14(6): 432-5.)

Effects of Andrographolide on Vascular Response

In a 2010 study, effects of andrographolide from *A. paniculata* on sexual functions, vascular reactivity and serum testosterone level in experimental animals were observed. The suspension of andrographolide in 5 per cent dimethyl sulfoxide (DMSO) was administered orally at the dose of 50 mg/kg to male ICR mice. The female mice involved in mating were made receptive by hormonal treatment. Administration of andrographolide significantly decreased the mounting latency at 120 and 180 min and increased mounting frequency at 180 min after treatment. Administration of 50 mg/kg andrographolide orally to male mice once daily for 2, 4, 6 or 8 weeks had no significant effects on sperm morphology and motility. At week 4, serum testosterone level in mice treated with andrographolide was significantly increased when compared to the control. (Sattayasai J *et al., Food and Chemical Toxicology*, 2010, Vol. 48(7): 1934-1938.)

Abortifacient and Non-teratogenic Effects

Studies in mice and rabbits suggest that the herb may have abortifacient activity. Conversely, no interruption of pregnancy, fetal resorption or decrease in the number of live offsprings was observed in pregnant rats after intragastric administration of an extract of the aerial parts at 2 g/kg bw during the first 9 days of gestation. Since potential antagonism exists between the herb and endogenous progesterone, the herb should not be used during pregnancy. (*WHO monographs on selected medicinal plants*, 2002, Vol. 2: 20.)

The effect of the powdered extract of *A. paniculata* (APE) leaves, an active ingredient of *Kan Jang* tablets [standardized for content of andrographolide (4.6 per cent) and 14-deoxo-andrographolide (2.3 per cent) content (total andrographolids—6.9 per cent)] on blood progesterone content in rats was studied. Peroral administration of APE during the first 19 days of pregnancy in doses of 200, 600, and 2000 mg/kg (*i.e.* doses 30, 90, and 300 fold higher than its daily therapeutic dose in humans) did not exhibit any effect on the elevated level of progesterone in the blood plasma of rats. The result indicates that in therapeutic dose, *A. paniculata* extract cannot induce progesterone-mediated termination of pregnancy. (Panossian A *et al., Phytomedicine*, 1999, 6: 157-161.)

Positive Effect of Low Doses

A small, short duration, phase 1 clinical trial in healthy men found no negative effects at doses 3 times the usual daily dose of *Kan Jang*. Instead, a positive trend in the number of spermatozoids, per cent active forms, and fertility indexes was found. Limitation of the study included the small number of participants (N=14), and short duration of the study (10 days), and the low doses tested (3 times the normal dose versus10-fold in animal studies). (*The Review of Natural Products*, Wolters Kluwer, 7th Edition, 2012: 79.)

Annona squamosa Linn.

Gandagaatra, Custard Apple

Anti-implantation Activity

The ethanolic extract of the seeds at a dose of 200 mg/kg *p.o.* on days 1 to 7 of pregnancy revealed 50 per cent anti-implantation activity in rats. The extracts at the doses of 50 and 100 mg/kg revealed 16.6 and 33.6 per cent reduction in pregnancy, respectively.

Depressant Effect on Central Nervous System

The ethanolic extract of seeds caused transient hypotension in anaesthetised cats with slight respiratory stimulation. Atropine completely blocked the effect of the extract on blood pressure. Depressant effects were observed on the auricular movements of open chest cat heart preparations *in situ*. The effect could be abolished by pretreatment by atropine.

An ethanolic extract of the defatted seeds produced a depressant effect on central nervous system, potentiated pentobarbitone hypnosis, showed anticonvulsant activity against electrically-induced convulsions and raised the pain threshold when tested by analgesiometer in rats. The ethanolic extract of the defatted seeds failed to inhibit rat paw edema induced by carageenin. It produced negative inotropic and chronotropic effects on perfused frog's heart which were unaltered by atropine pretreatment. (*Reviews on Indian Medicinal Plants*, ICMR, 2004, Vol. 2: 361.)

Cytotoxicity

Etanolic extract of the bark yielded the bioactive acetogenins, squamotacin and molvizarin. Squamotacin showed cytotoxic selectivity for the human prostate tumour cell line (PC-3) with a potency of over 100 million times that of Adriamycin. (Hopp *et al., J Nat Prod*, 1996, 59: 97.)

Toxicity of Acetogenins

Over 220 types of acetogenins have been identified from a related species, *Annona muricata*; annonacin is the most predominant one. It crosses the blood-brain barrier. Annonacin and the alkaloids reticuline and coreximine have been evaluated for toxic effects on rat dopaminergic neurons *in vitro*. The mechanisms of action is unclear but are suspected to involve the inhibition of dopamine uptake as well as effects on neuronal energy production and mitochondrial respiration. Nigral and striatal degeneration in rats has been demonstrated and alkaloid-induced cell death was also observed. (*www.drugs.com)*

Caution

One report estimates that the amount of annonacin ingested by one adult eating one fruit of *Annona muricata* daily for one year is comparable with the intravenous dose used to induce brain lesions in rats. One fruit contains approximately annonacin 15 mg, and a can of commercial nectar contains 36 mg. (*The Reviews of Natural Products,* 7th Edn. 2012: 1514.)

Anthocephalus cadamba Miq.

Kadamba, Burflower Tree

Abortifacient Potential

A. cadamba has been known to possess abortifacient potential in ethnobotanical literature, but has not been validated scientifically.

In a study, the methanolic extract of the stem bark was prepared and tested for abortifacient, estrogenic and uterotrophic activity. Pregnant Swiss albino mice were randomized into 5 groups (1-5.) Group 1 (negative control) received 0.2 per cent w/v agar, group 2-4 (received extract at the dose of 500, 1000 and 1500 mg/kg b.w.) and group 5 received mifepristone at a dose of 5.86 mg/kg b.w. respectively, by oral route from 10th to 18th day post coitum daily, and various parameters recorded. The uterotrophic bioassay was performed in bilaterally ovariectomized mice dosed from 9th to 15th day of ovariectomy and change in uterotrophic parameters was observed.

Preliminary phytochemical screening revealed presence of glycosides, alkaloids, steroids, saponins, triterpenoids, flavonoids and tannins. No signs of clinical toxicity were observed at any time during the period of treatment. The extract significantly reduced ($P<0.05$) the number of live fetus, weight and survival ratio of the fetus, number of corpora lutea, progesterone, estradiol and luteinizing hormone whereas the number of dead fetus, number of mice that aborted, percentage vaginal opening and post-implantation loss increased significantly ($P<0.05$.) The estrogenicity experiments showed increase in uterine weight ($P<0.05$), ballooning of uterus, uterine glucose ($P<0.05$) and ALP ($P<0.001$) in extract treated group dose dependently. In addition, the extract also induced vaginal bleeding preceding parturition.

This study substantiated the abortifacient potential of the methanolic extract of *A. cadamba* stem bark. The activity was more marked in 1000 and 1500 mg/kg b.w. of the extract and was comparable to that of mifepristone. The mechanism of abortion could possibly be through changes in the uterine mileu, altered hormone levels, luteolysis and partly, estrogenicity. (Shaikh MV *et al.,* Abortifacient potential of methanolic extract of *Anthocephalus cadamba* stem bark in mice, *J Ethnopharmacol,* 2015 Sep 15,173: 313-317.)

Sedative and Antiepileptic Effects

In a study, angiotensin converting enzyme (ACE) was tested at three doses *viz.,* 100, 200 and 400 mg/kg p.o. Ketamine-induced sleeping time model was used to test the sedative property of the extract where, onset and duration of sleep were observed. A paradigm of anticonvulsant models pentylenetetrazole (PTZ), isoniazid (INH) and maximal electroshock (MES)-induced seizures) were used to evaluate its protective effect against absence and generalized types of seizures. Onset of clonic convulsions, tonic extension and time of death were observed in PTZ- and INH-induced seizure models. In MES model, duration of tonic hind leg extension and onset of stupor were observed. ACE showed significant increase in ketamine induced sleeping time. It also exhibited significant increase ($P<0.05$, 0.01 and 0.001) in latency to clonic convulsion, tonic extension and time of death in PTZ and INH models at all tested doses, whereas in the MES model, the lower dose was found to be effective when compared with the higher doses (200 and 400 mg/kg, p.o.).

A. cadamba bark contains alkaloids like cadambine and its derivatives, saponins, glycosides, triterpinoids, cadambagic acid, quinovic acid and beta-sitosterol. Several reports suggest that alkaloids, triterpenic steroids and flavonoids have potent antiepileptic effect in various seizure models. In addition to this, saponins have also been able to modulate the neurotransmitter levels in the brain and to possess potent anti-convulsant activity. Therefore, the presence of such compounds in the extract may be responsible for the sedative and anticonvulsant activities. (Pandian Nagakannan *et al., Indian Journal of Pharmacology,* Short Communication, Sedative and antiepileptic effects of *Anthocephalus cadamba.* in mice and rats, November-December, 2011, 43(6): 699-702.)

Apium graveolens Linn.

Ajmoda, Celery

Thyroxine Interacts with Celery Seed Tablets: Case Reports

The first case involved a 55-year-old woman who, after considerable monitoring, had finally been stabilised on a daily dose of thyroxine 100 microgram. A month later, her doctor found that her T_4 levels were low again and her dose was doubled. The patient then remembered that in the past month she had also started taking celery seed tablets for osteoarthritis. Suspecting a potential interaction, she ceased the celery seed tablets without increasing the thyroxine dose as the doctor had advised. Next time her thyroxine levels were checked they were found within the normal range. She tried recommencing celery seed a month later but after a week she felt lethargic, bloated and had dry skin. When she stopped the celery seed tablets, she reported that her 'general energy levels improved.'

A second report was received from a 49-year-old woman who had taken thyroxine for many years. When her T_4 became extremely low her doctor suspected that she had not been taking her tablets. The patient argued that she had taken her thyroxine, but she had recently commenced taking celery seed tablets to treat arthritis. She ceased the celery seed tablets and one month later her thyroxine levels had returned to within the normal range. (Geraldine Moses, *Aust Prescr*, Jan 1 2001, 24: 6-7.)

When reference was made to these case studies in an article in a Queensland newspaper, the Queensland Medication Helpline received a flood of calls about similar experiences. A total of 10 cases are now on file. A pharmacokinetic study of the T_4-celery interaction is under consideration by the Mater Hospital Pharmacy Services, Therapeutic Advisory Service.

Areca catechu Linn.

Puuga, Betel Nut

Carcinogenic, Genotoxic and Cardiotoxic Effects

In India, role of Areca nut consumption in oral cancer using cytological studies to assess the possible genomic damage caused by its consumption (without tobacco, and not as a component of betal-quid) was studied. The analysis showed statistically significant increase in the frequencies of SCEs (Sister chromatid exchange) and chromosome aberrations in peripheral blood lymphocytes and the percentage of micronucleated cells in exfoliated cells of buccal mucosa among the group of chewers (normal, with submucous fibrosis, and with oral cancer) when compared with those of (healthy, non-chewing) controls. (*Reviews on Indian Medicinal Plants*, ICMR, New Delhi, Vol.3, 2004: 16-17.)

The aqueous extract of Areca nut as well as arcoline showed *in vitro* genotoxic effect in Chinese hamster ovary cells. The chromosome damage was more severe on treating the cells with low concentrations and for longer duration, which mimics the effects of chronic Areca nut consumption. (*The Wealth of India*, First Supplement series, Vol. 1: 86.)

A clinical trial investigated the association between betel nut chewing and cardiovascular disease (CVD.) A baseline cohort of 56,116 male participants were recruited from 4 nationwide health screening centers in Taiwan in 1998 and 1999. There were 1549 deaths during the follow-up period, 309 of which were due to CVD. Current and former betel nut chewers had a higher risk of CVD mortality than did current and former smokers. (Lin WY *et al.,* Betel nut chewing is associated with increased risk of cardiovascular disease and all-cause mortality in Taiwanese men, *Am J Clin Nutr*, 2008 May, 87(5): 1204-11.)

Artemisia absinthium Linn.

Damanaka, Wormwood

Toxicity of Thujone

People who overuse wormwood or take wormwood essential oil, which contains higher levels of thujone than wormwood tea or tincture, may have seizures or convulsions, or experience hallucinations. In one German study, researchers gave 20 patients suffering from Crohn's disease an herbal preparation that contained low concentration of thujone. The study reported no serious side effects because a relatively low concentration of thujone. (*livestrong.com*)

In a 13-week dose-toxicity study, convulsions were observed in rats given thujone in concentrations as low as 25 mg/kg/day. An increase in mortality was shown in rats given 50 mg/kg/day. Other studies document a dose of 120 mg/kg as fatal, including a subcutaneous median lethal dose of thujone in mice as 134 mg/kg.

Wormwood is classified as an unsafe herb by FDA because of the neurotoxic potential of thusone and its derivatives. (*The Review of Natural Products*, Wolters Kluwer, 7th Edn, 2012: 1715.)

Sedative Effect

A dry 80 per cent -methanolic wormwood extract, orally administered to mice at 500 mg/kg) caused significant prolonged pentobarbital-induced sleeping time from 81 minutes to 117 minutes ($P < 0.05$.) (ESCOP, Second Edn, 2003: 4.)

Contraindications

Theoretically, wormwood may be contraindicated in patients with an underlying defect with hepatic heme synthesis because thujone is a porphyrogenic terpenoid. Thujone is metabolised via the cytochrome P450 enzyme system with involvement of the specific enzymes CYP2A6, CYP3A4 and CYP2B6. Therefore, there is a theoretical risk of interactions with drugs and other herbs metabolised via this system. (cam-cancer.org)

Artocarpus heterophyllus Lam.

Panasa, Jackfruit

Effect of Sedative Activity of Roasted Seeds on Sexual Behaviour

According to a medicinal plants text of Sri Lanka, roasted seeds of *Artocarpus heterophyllus* has aphrodisiac activity. However, some reproductively active young men in rural areas of Sri Lanka claim that consumption of these seeds few hours prior to coitus disrupts sexual function. Because of these two conflicting claims, it was thought useful to scientifically investigate the effects of *A. heterophyllus* seeds on male sexual function and fertility. This was done using a seed suspension in 1 per cent methylcellulose (SS) in rats. In a sexual behaviour study using receptive female rats, an oral administration of 500 mg/kg dose of SS markedly inhibited libido, sexual arousal, sexual vigour and sexual performance within 2 hr. Further, the treatment induced a mild erectile dysfunction.

These antimasculine effects on sexual function were not evident 6 hr post treatment indicating rapid onset and offset of action. Further, these actions on the sexual behaviour was not due to general toxicity, liver toxicity, stress or reduction in blood testosterone level but due to marked sedative activity. In a mating study, SS failed to alter ejaculating competence and fertility. These results suggest that *A. heterophyllus* seeds do not have aphrodisiac action, at least, in rats. (W.D. Ratnasooriya, J.R. Jayakody. *Artocarpus heterophyllus* seeds inhibit sexual competence but not fertility of male rats, *Indian J Exp Biol, 2002,* 40(3): 304-308.)

Asparagus racemosus Willd.

Shataavari, Indian Asparagus

Teratogenic Effect

With the methanolic extract of *Asparagus racemosus* (50-2000 mg/kg or 28 days treatment with 100-1000 mg/kg) no acute or subacute toxicity in adult mice, in terms of mortality, was observed. However, pre- and post-natal studies did indicate some teratogenic effect. The methanolic extract, 100 mg/kg/day for 60 days, showed teratological disorders in terms of increased resorption of foetuses, gross malformations *e.g.* swelling in legs and intrauterine growth retardation with a small placental size in Charles Foster rats. Pups born to mother exposed to methanolic extract of *Asparagus racemosus* for full duration of gestation showed evidence of higher rate of resorption and therefore smaller litter size. The live pup showed significant decrease in body weight and length and delay of various developmental parameters when compared to respective control groups. (RK Goel *et al.*, *Indian J Experimental Biology*, July 2006, Vol 44: 570-573.)

Inhibitory Effects on the Digestive Enzymes?

The roots are reported to show inhibitory effects on the digestive enzymes, lipase and trypsin, and may lead to stoppage of degradation of food material in the intestinal tract. (Vijaya and Vasudevan, *Indian Drugs*, 1994, 31: 205. Cited in *The Wealth of India*, First Supplement series,Vol.1: 102.)

In another study, *Asparagus racemosus* was compared with a modern drug, metoclopramide, which is used in dyspepsia to reduce gastric emptying time. Metoclopramide and *Asparagus racemosus* did not differ significantly in their effects. (Dalvi SS *et al.,* Department of Pharmacology, Seth G. S. Medical College, Mumbai. *Postgrad Med J*, 1990, 36 (2): 91-96.)

Azadirachita indica A. Juss.

Nimba, Neem

Cardiovascular Effect

Ethanol extract of Neem leaf induces dose-dependent hypotensive action in rats, but bradycardia, as well as cardiac arrhythmia, was also observed.

Spermicidal Activity of Leaf Extract

Salanin and sodium nimbidinate compounds are spermicidal to rats and human spermatozoa. Antiandrogenic properties have been demonstrated in rats and oral neem extract decreased total sperm count and sperm motility in rats; additionally, increased the proportion of abnormal sperm.(*The Review of Natural Products*, Wolters Kluwer, 7ᵗʰ Edition, 2012: 1155.)

Effect on Thyroid Hormones

In a study, Neem leaf extract was orally administered in two different doses (40 mg and 100 mg kg(-1)day(-1)for 20 days) in male mice. The extract exhibited differential effects. While the higher dose decreased serum tri-iodothyonine (T_3) and increased serum thyroxine (T_4)) concentrations, no significant alterations of levels were observed in the lower dose group, indicating that the high concentrations of neem extract can be inhibitory to thyroid function, particularly in the conversion of T_4 to T_3, the major source of T_3 generation. A concomitant increase in hepatic lipid peroxidation (LPO) and a decrease in glucose-6-phosphatase (G-6-Pase) activity in the higher dosed group also indicated the adverse effect of neem extract despite an enhancement in the activities of two defensive enzymes, superoxide dismutase (SOD) and catalase (CAT.) Higher concentration of neem extract may not be safe with respect to thyroid function and lipid peroxidation. (Panda S, Kar A, *Pharmacol Res*, 2000 Apr, 41(4): 419-422.)

Bacopa monnieri (Linn.) Penn.
Braahmi, Thyme-leaved Gratiola

Effects on Regulation of Hormones

Relative importance of *Bacopa monnieri* (200 mg/kg), *Aegle marmelos* (1 g/kg) and *Aloe vera* (125 mg/kg) leaf extracts in the regulation of thyroid hormone concentrations in male mice was investigated. While serum levels of both T_3 and T_4 were inhibited by *A. vera*, *A. marmelos* extract could decrease only T_3 concentration. On the other hand, T_4 concentration was increased by *B. monnieri* extract suggesting its thyroid-stimulating role. When the relative potency of each plant extract was calculated in terms of percent increase or decrease in thyroid hormones, as compared to the control value, the decrease in T_3 concentration by *A. marmelos* was about 62 per cent indicating its possible use in the regulation of hyperthyroidism. *B. monnieri* could increase T_4 concentration by 41 per cent without enhancing hepatic lipid peroxidation (LPO) suggesting that it can be used as a thyroid-stimulating drug. In fact, hepatic LPO was decreased and superoxide dismutase (SOD) and catalase (CAT) activities were increased by *B. monnieri* and *A. marmelos* leaf extracts showing their antiperoxidative role. It is thus suggested that *A. marmelos* and *A. vera* may be used in the regulation of hyperthyroidism, while *B. monnieri* in hypothyroidism. (Kar A, Panda S, Bharti S, *J Ethanopharmacol*, 01Jul, 2002, 81(2): 281-285. PMID 12065164.)

Effect on Fertility

A study evaluated the effect of *Bacopa monnieri* on fertility of male laboratory mouse. Mice of the Parkes (P) strain were orally administered Brahmi (250 mg/kg body weight/day, for 28 and 56 days), and effect of the treatment on reproductive organs and fertility was investigated. Recovery and toxicological studies were also carried out. The treatment caused reduction in motility, viability, morphology, and number of spermatozoa in cauda epididymidis. Histologically, testes in mice treated with the plant extract showed alterations in the seminiferous tubules, and the alterations included intraepithelial vacuolation, loosening of germinal epithelium, exfoliation of germ cells and occurrence of giant cells. In severe cases, the tubules were lined by only sertoli cells, spermatogonia and spermatocytes. Significant reductions were also noted in height of the germinal epithelium and diameter of the seminiferous tubules in Brahmi-treated mice compared to controls. Epididymis in treated males showed slight alterations in histological appearance. The treatment had no effect on levels of testosterone, alanine aminotransferase, aspartate aminotransferase and creatinine in blood serum, hematological parameters and on liver and kidney histoarchitecture. In Brahmi-treated males, libido remained unaffected, but fertility was notably suppressed. The alterations caused in the above reproductive endpoints by the plant extract were reversible, and by 56 days of treatment withdrawal, the parameters recovered to control levels. (Singh A, Singh SK, Department of Zoology, Banaras Hindu University, Varanasi, India, Evaluation of antifertility potential of Brahmi in male mouse, *Contraception*, 2009 Jan; 79(1): 71-9.)

Caution

Centella asiatica, also known as Brahmi, should not be used as a substitute of *Bacopa monnieri*. Brahmi promotes fertility and sustains implantation and pregnancy, while *Centella asiatica* tend to do the opposite. (Sivarajan and Indira Balachandran, *Ayurvedic Drugs and their Plant Sources*, Oxford and IBH Publishing Company, !984: 97.)

Balanites aegyptiaca (L.) Delile.
Ingudi, Desert Date

Toxicity on Spermatogenic Elements

Chronic administration of the ethanolic extract of the fruit pulp in a dose of 35 mg/kg bw daily for 60 days in dogs, resulted in a mass atrophy of the spermatogenic elements, reduced the total proteins, salic acid and glycogen contents of the testes with increased testicular cholesterol and alkaline phosphatase in dogs.

The aqueous alcoholic seed kernel extract showed anti-ovulatory, anti-oestrus and sperm motility inhibitory action without local irritation in rabbits and rats. (*Reviews on Indian Medicinal Plants*, ICMR, New Delhi, Vol. 4, 2004: 30.)

The plant is reported to be a potential source of diosgenin (used in oral contraceptives.) The diosgenin content of the fruit varies from 0.3 to 3.8 per cent. Sperms became sluggish on contact with the plant extract and then became immobile within 30 s; the effect was concentration-related. Protracted administration of the fruit pulp extract produced hyperglycaemia-induced testicular dysfunction in dogs. (Khare CP, *Indian Medicinal Plants: An Illustrated Dictionary*, Springer, 2007: 7.)

Dose-dependent Effect in Jaundice

The aqueous extract of *Balanites aegyptiaca* bark, which is used in Sudanese folk medicine in the treatment of jaundice, was without effect when studied on rabbit intestine, rabbit aortic strip, rat stomach strip, rat uterus and rat phrenic nerve–diaphragm in a dose up to 10 mg/mL gut bath. In a larger dose (25 mg) the aqueous extract to biliary ductligated rats, showed a dosedependent significant decrease in serum bilirubin level. The chronic and subchronic toxicity investigations indicate the safety of the aqueous extract at 25 mg dose level. (A.H. Mohamed *et al.*, *Phytotherapy Research*, Aug/Sept 1999, 13(5): 439-441.)

Bauhinia variegata Linn.
Kaanchnaara, Mountain Ebony

Chronic Toxicity of Stem Bark

Stem bark of Kanchanara (*Bauhinia variegata* L., Family-Caesalpiniaceae) is used in Ayurvedic system of medicine, either as a single drug or as ingredient of compound formulations. The present study was carried out to evaluate the toxicity of stem bark where its powder suspension was administered in Therapeutic Equivalent Dose (TED) (350 mg/kg/day) in TED group and five fold of TED (1800 mg/kg/day) in TED× 5 group for sixty days in albino rats. Control group received distilled water. Parameters like body weight, weight of important organs, biochemical, hematological were studied along with histopathology of vital organs. Kanchanara at both the dose levels significantly increased the weight of spleen and thymus. Decreased HDL cholesterol and direct bilirubin were observed in both the treated group while decreased blood urea was observed in TED×5 group. Significant increase in platelet count and significant decrease in haemoglobin and lymphocyte count were observed in TED×5 treated group. In histopathological study, destruction of epithelial layer of stomach was observed in majority of the sections in TED×5 dose groups compared to control group. Thus, toxicity profile obtained from the present study showed that *B. variegata* bark is likely to produce toxic effect when administered in a five times dose in powder form. (Rasika Kolhe *et al.,* Chronic toxicity study of stem bark powder of *Bauhinia variegata* in albino rats, *Indian Journal of Natural Products and Resources*, Oct 2014, 5(3): 244-248. ResearchGate.)

Effect on Thyroid Hormones

Daily administration of *Withania somnifera* root extract (1.4 g/kg body wt.) and *Bauhinia purpurea* bark extract (2.5 mg/kg body wt.), for 20 days, were investigated for thyroid function in female mice. While serum triiodothyronine (T_3) and thyroxine (T_4) concentrations were increased significantly by *Bauhinia*, *Withania* could enhance only serum T_4 concentration. Both the plant extracts showed an increase in hepatic glucose-6-phosphatase (G-6-Pase) activity and antiperoxidative effects as indicated either by a decrease in hepatic lipid peroxidation (LPO) and/or by an increase in the activity of antioxidant enzyme(s). It appears that these plant extracts are capable of stimulating thyroid function in female mice. (S. Panda, A. Kar, *J Ethnopharmacol.* Nov. 1999, 67(2): 233-239.)

Berberis aristata DC.

Daaruharidra, Indian Barberry

Berberine occurs chiefly in *B. aristata* var. *aristata* and other species of *Berberis*. A number of studies have shown berberine to be effective against diarrheas caused by a number of different types of parasites and infectious organisms including *E. coli* (traveller's diarrhea), *Shigella dysenteriae* (shigellosis), *Salmonella paratyphi* (food poisoning), *B. klebsiella*, *Giardia lamblia* (giardiasis), *Entamoeba histolytica* (amebiasis), and *Vibrio cholerae* (cholera.)

In Ayurvedic medicine, Rasaut, an extract prepared from the wood or root of Berberis species, is used. Out of fifteen samples collected from the market, 7 did not contain alkaloids or any traces of it. The variation in the alkaloidal content in the rest of the samples was from 1.67 to 4.26 per cent. (Grewal and Kochhar, 1940. *Cf. Reviews on Indian Medicinal Plants*, ICMR, Vol. 4: 159.) The root and stem bark of *Berberis* species contain alkaloids 5 and 4.2 per cent, respectively, calculated as berberine.

The Effect of Berberine on Dopamine

Post-traumatic stress disorder (PTSD) is a trauma-induced psychiatric disorder characterized by impaired fear extermination, hyperarousal, anxiety, depression, and amnesic symptoms that may involve the release of monoamines in the fear circuit. A recent study measured several anxiety-related behavioural responses to examine the effects of berberine on symptoms of anxiety in rats after single prolonged stress (SPS) exposure, and to determine if berberine reversed the dopamine dysfunction. Rats received berberine (10, 20, or 30 mg/kg, intraperitoneally, once daily) for 14 days after SPS exposure. Berberine administration significantly increased the time spent in the open arms and reduced grooming behaviour during the elevated plus maze test, and increased the time spent in the central zone and the number of central zone crossings in the open field test. Berberine restored neurochemical abnormalities and the SPS-induced decrease in dopamine tissue levels in the hippocampus and striatum. The increased dopamine concentration during berberine treatment may partly be attributed to mRNA expression of tyrosine hydroxylase and the dopamine transporter in the hippocampus, while berberine exerted no significant effects on vesicular monoamine transporter mRNA expression in the hippocampus of rats with PTSD. These results suggest that berberine had anxiolytic-like effects on behavioural and biochemical measures associated with anxiety. These findings support a role for reduced anxiety altered DAergic transmission and reduced anxiety in rats with PTSD. Thus, berberine may be a useful agent to treat or alleviate psychiatric disorders like those observed in patients with PTSD. (Lee B *et al.,* Berberine alleviates symptoms of anxiety by enhancing dopamine expression in rats with post-traumatic stress disorder, *Korean J Physiol Pharmacol*, 2018 Mar, 22(2): 183-192. Published online 2018 Feb 23.)

Effect on Drug Withdrawal

Berberine petreatment prior to every morphine treatment reduced depression- and anxiety-like symptoms strongly associated with morphine discontinuation, probably by modulating hypothalamic corticotrophin-releasing factor and the noradrenergic system in the CNS. Therefore, berberine may be a useful compound in the development of alternative medicines for treating morphine withdrawal- related symptoms, such as depression and anxiety. (Lee B *et al.*, Effect of berberine on depression- and anxiety-like behaviors and activation of the noradrenergic system induced by development of morphine dependence in rats, *Korean J Physiol Pharmacol*, 2012, 16: 379-386.)

Effect on Drug Abuse

Recently, berberine was reported to inhibit the rewarding effects of drugs of abuse such as cocaine, morphine, and nicotine. Berberine is also demonstrated to modulate the activity of several neurotransmitter systems like, dopamine, nitric oxide, serotonin, and NMDA, which are implicated in rewarding effects of ethanol.

In an experiment, the effect of berberine as studied on locomotor sensitization, conditioned place preference (CPP), and ethanol drinking preference in mice. The results revealed that acute administration of berberine (2.5, 5, and 10 mg/kg, *i.p.*), dose dependently, reduced locomotor stimulant effect of acute ethanol and expression of sensitization to locomotor stimulant effect of ethanol. Further, pretreatment with berberine (2.5, 5, and 10 mg/kg, i.p.) prior to each dose of ethanol, blocked the development as well as expression of sensitization to locomotor stimulant effect of ethanol. In another set of experiment, treatment with berberine (5 and 10 mg/kg, i.p.) reduced the induction and expression of ethanol-induced CPP in mice. In addition, berberine in these doses also reduced preference to ethanol drinking over water, but did not alter the general reward. The results of the study revealed that berberine attenuates ethanol-induced rewarding effects in mice and that could be attributed to its neuro-modulatory action. (Pravinkuimar Bhutada *et al.*, Inhibitory effect of berberine on the motivational effects of ethanol in mice, *Progress in Neuro-Psychopharmacology and Biological Psychiatry*, December 2010, 34(8): 1472-1479. ResearchGate.)

Interactions with Drugs

Berberine may interfere with the absorption of tetracycline and related antibiotics. Berberine also produces significant inhibition of CYP3A enzymes in humans. Because most drugs are metabolized by these enzymes, berberine may decrease the clearance of many medications thereby potentiating their effect.

Berberine-containing plants may enhance the effects of oral hypoglycemic drugs used in the treatment of type 2 diabetes through its multitude of anti-diabetic effects. People on oral hypoglycemic drugs should monitor blood glucose levels if taking berberine and adjust their dosage of their medications under medical advice. (doctormurray.com)

Comments in German Commission Monograph, March 2,1989

German Commission E placed *Berberis vulgaris* bark of the aboveground parts as well as that of the root) among unapproved herbs, and listed the risks observed with purified alkaloid berberine, not parts of the plant. With intake of more than 0.5 g of berberine, following symptoms are described:

In small doses berberine stimulates the nervous system, while higher doses lead to severe dyspnea and spasm ending in lethal primary paralysis of the respiratory system. Lethal doses also cause hemorrhagic nephritis. Death due to respiratory paralysis occurred in anaesthetized cats and dogs at 25 mg/kg. In addition, a noticeable inhibition of the heart action was observed. The LD_{50} for berberine sulphate in mice is 24.3 mg/kg in intraperitoneal application. No reports of poisoning with *Berberis vulgaris* herb are known. (*The Complete German Commission E Monographs*, Blumenthal *et al.*, *American Botanical Council*, 1998: 310.)

A Forecast

Berberine should be a much more popular herbal remedy than curcumin. Not to discredit curcumin, but the existing clinical research with berberine is just so much stronger. It is not even close to berberine. (Dr Michael Murray, One of the world's leading authorities on Natural Medicine.)

Brassica alba Boiss

Sarshapa Gaura, White Mustard

Effect on Thyroid Hormones

Isothiocyanates such as those present in mustard have been implicated in endemic goiter (hypothyroidism with thyroid enlargement.) They also have been reported to produce goiter in experimental animals. Allyl isothiocyanate and butenyl isothiocyanate are goitrogenic but are strong antimutagens.

A study was aimed to assess the possible goitergentic of amiodarone and mustard oil if they are given together by measuring thyroid function tests including (TSH, T_3 and T_4), weight of thyroid gland. Heart enzyme CK and the weight of the heart as well as ECG readings, in addition, the liver function enzymes (SGOT and SGPT) all these were measured. 30 adult male rabbits (weight range 1250-2000 g and age-9-12 months) were divided randomly to four groups each contain 8 animals except control group contained 6 animals. Second, third and fourth groups were given mustard oil (2 g/d), amiodarome (8 mg/kg) and combined (mustard oil 2 g/d + amiodarome 8 mg/kg) intragastric therapy respectively for 2 weeks. First group was kept as control. Results: Combination of mustard oil and amiodarone was shown significant increase in thyroid hormones (T_3 and T_4.) This effect was supported by significant increased in the thyroid gland weight and reduction in body weight, reduced the liver function enzyme especially SGOT.

This study showed that mustard oil may increase the goiterogenic effect of amiodarone if they are administered concomitantly for long time. On the other hand this oil has hepato- and cardioprotective effect and can ameliorate the toxic effects of amiodarone on liver and heart. (Doa A Anwar Ibrahim, The impact of concomitant administration of antiarrhythmic agent (amiodarone) with mustard oil on thyroid gland in experimental animals, *International Journal of Scientific Research*, Volume II, Issue VIII, August 2013.)

Butea monosperma (Lam.) Taub.

Palaasha, Flame-of-the-forest

Toxic Effects of Seeds

In a study, toxic effects of powder of seeds of *Butea monosperma* were evaluated for a period of 3 months in albino rats. Control group received distilled water. The powder suspension was orally given to the treated group at a dose of 800 mg/kg/day for 90 days. Parameters like body weight, weight of important organs, biochemical, hematological parameters, bone marrow cytology and histopathology of vital organs were studied. Test drug administration did not affect the body weight, organ weight and bone marrow cytology to a significant extent. Among the 18 haematological parameters studied, significant changes were observed in three parameters, namely, significant decrease in haemoglobin content, red blood cell count and haematocrit. Of 16 biochemical parameters studied, significant changes were observed in 5 parameters, namely, decrease in total protein, albumin, bilirubin and significant increase in very low density lipoprotein and triglyceride. The histopathology of 18 organs revealed changes, such as fatty changes, glomerular congestion and tubular haemorrhage in the kidneys, decrease in the cellularity of the spleen, epithelial disruption in jejunum, decrease in spermatogenesis in the testis, epithelial proliferation in ventral prostate and decrease in epithelial proliferation in the uterus. Thus, toxicity profile showed that *B. monosperma* seeds are likely to produce toxic effect when administered in a powder form. (Shilpa Donga *et al.*, Chronic toxicity study of *Butea monosperma* (Linn.) Kuntze seeds in albino rats. *Ayu.* 2011 Jan-Mar; 32(1): 120–125.)

Anti-implantation Activity

Butin, which is isolated from the seeds of *Butea monosperma*, administered orally to adult female rats at the doses of 5, 10 and 20 mg/rat from day 1 to day 5 of pregnancy showed anti-implantation activity in 40 per cent, 70 per cent and 90 per cent of the treated animals, respectively. At lower doses, there was a dose-dependent termination of pregnancy and reduction in the number of implantation sites. In ovariectomized young female rats, the butin exhibited estrogenic activity at comparable anticonceptive doses, but was devoid of anti-estrogenic activity. Butin is a weak estrogen, a significant uterotrophic effect was discerned even at 1/20th the anticonceptive dose. (S.K. Bhargava. Estrogenic and postcoital anticonceptive activity in rats of butin isolated from *Butea monosperma* seed. *J Ethnopharmacol*, 1986, 18(1): 95-101.)

Antiesterogenic and Antifertility Activity

Methanolic extracts of *Butea monosperma* exhibited effect on uterotropic and uterine peroxidase activities in ovariectomized rats and determined estrogenic/antiestrogenic potential of antifertility substances using rat uterine peroxidase assay. Alcoholic extract of flowers has also been reported to exhibit antiestrogenic and antifertility activities. Butin isolated from the flowers show both male and female contraceptive properties. (S.K Bhargava, *Fitoterapia*, 1988, 59: 163.)

Thyroid Inhibitory, Antiperoxidative and Hypoglycemic Effects

Stigmasterol, isolated from the bark of *Butea monosperma* was evaluated for its thyroid hormone and glucose regulatory efficacy in mice by administrating 2.6 mg/kg/d for 20 days which reduced serum triiodothyronine (T_3), thyroxin (T_4) and glucose concentrations as well as the activity of hepatic glucose-6-phophatase (G-6-Pase) with a concomitant increase in insulin indicating its thyroid inhibiting and hypoglycemic properties. A decrease in the hepatic lipid peroxidation (LPO) and an increase in the activities of catalase (CAT), superoxide dismutase (SOD) and glutathione (GSH) suggested its antioxidative potential. The highest concentration tested (5.2 mg/kg) evoked pro-oxidative activity. (S. Panda *et al.*, Thyroid inhibitory, antiperoxidative and hypoglycemic effects of stigmasterol isolated from *Butea monosperma. Fitoterapia*, 2009, 80(2): 123-126.)

Calotropis procera (Ait.) R.Br.

Alarka, Milk-weed

Cardiac and Testicular Toxicity

In a study, the toxic effect of latex and ethanolic leaf extract of *Calotropis procera*, in comparison to abamectin, albino rats were separately administered 1/20 and 1/10 of LD_{50} of *C. procera* latex, ethanolic *C. procera* leaf extract and abamectin respectively by oral gavage for 4 and 8 weeks. *C. procera* latex and leaf extract as well as abamectin markedly elevated the activities of serum CK-MB, AST and LDH at the two tested periods in a dose dependent manner. Lipid peroxidation was significantly increased while GSH level and GPx, GST and SOD activities were significantly depleted in heart and testis of all treated rats. All treatments also induced a marked increase in serum TNF-α and decrease in serum IL-4, testosterone, FSH and LH levels in a dose dependent manner. The latex seemed to be more effective in deteriorating the testicular function and sex hormones' levels while the ethanolic leaf extract produced more deleterious effects on oxidative stress and antioxidant defense system in both heart and testis. The normal histological architecture and integrity of the heart and testis were perturbed after treatments and the severity of lesions, which include odema, inflammatory cell infiltration, necrosis and degeneration, is dose and time dependent. In conclusion, the findings of this study indicated that *C. procera* latex and ethanolic extract of leaves could induce marked toxicity in heart and testis and these toxic effects may be more or less similar to those of abamectin. The cardiotoxicity and testicular toxicity may be mediated via stimulation of inflammation, increased oxidative stress and suppression of antioxidant defense system. (Osama M. Ahmed *et al., Springerplus*, 2016, 5(1): 1644. PMCID: PMC5033794.)

Degeneration of Spermatogonia and Anti-implantation Activity

Oral administration of crude extract of flower and root (20 mg alternate day/ animal for 30 days) to Indian adult desert male gerbils revealed testicular necrosis, degeneration of spermatogonia, spermatocytes and sertoli cells.

Calotropin, when administered to gebrils (25 mg/kg bw) and rabbits (25 mg/kg bw) once a day for 30 days inhibited the process of spermatogenesis. The population of spermatids was depleted by 65 and 94 per cent in gebrils and rabbits, respectively. The seminiferous tubules and Leydig cell nuclei diameters were reduced in both the species.

Calotropin at a dose of 25 mg/kg bw also exhibited abortifacient activity in pregnant rats, The 50 per cent ethanolic extract of the root in doses of 30 and 50 mg/ daily showed 83.4 and 75 per cent anti-implantation activity in female rats respectively.

The ethanolic extract of the root at a dose level 250 mg/kg exhibited 100 per cent anti-implantation and uterotropic activity in albino rats.

Tissue Damage

Oral administration of crude extract of the root and flowers (20 mg/alternate day/animal for 30 days) produced a decrease in protein, RNA and acid contents in tissues and increase in acid phosphatase, alkaline phosphatase, serum transasminase, cholesterol and total lipid contents. Liver was damaged as revealed by tumour-like structure visible in the liver lobes.

Toxicity on Liver, Kidney and Alimentary Canal

The toxicity study was conducted for latex-fed to black rats through bait prepared from wheat flour, groundnut oil and sugar at the concentration of 5, 7.5 and 10 per cent (w/W) for 10 days. Rats exhibited passivity, sluggishness, sedation, dyspnea, weakness, reduction in weight, anorexia, diarrhea, haematuria, bleeding from nose, eyes and anus, eye lens opacity, mild tetanic, convulsions, collapse and death. Motalities were 56.75, 68.75 and 87.5 per cent with the respective doses. Changes were noted in liver, kidney and alimentary canal. (*Reviews on Indian Medicinal Plants*, ICMR, 2005, Vol. 5: 187-188.)

Ocular Toxicity by Latex

Ocular toxicity, following accidental inoculation of latex of *Calotropis procera* (in 29 eyes between January 2003 and December 2006, has been reported, All patients presented with sudden painless dimness of vision with photophobia. Twenty-five (86 per cent) patients had initial visual acuity of less than 20/60. All eyes had conjunctival congestion and mild to severe corneal edema with Descemet's folds. Three (10 per cent) eyes had an epithelial defect, nine (31 per cent) had iridocyclitis, and seven (24 per cent) had associated secondary glaucoma. After treatment with topical corticosteroids, antiglaucoma agents, cycloplegics, hypertonic saline and tears supplements, 27 (93 per cent) eyes recovered completely within 3–14 days. After three months, 17 (74 per cent) out of 23 eyes showed a significant low endothelial cell count compared to the normal fellow eye ($P < 0.001$). (Samar K Basak *et al.*, Ocular toxicity by latex of *Calotropis procera*, *Indian J Ophthalmol*, 2009 May-Jun; 57(3): 232–234.)

Cannabis sativa Linn.

Ganja, Marijuana, Indian Hemp

Biological Effects on 'Sadhus' Addicted to Cannabis (Ganja)

Eleven cannabis users belonging to the same group of *'sadhus'* (17-50 years) and 7 non user *'sadhus'* (19-55 years) as control were included in a study. Each individual consumed 1.25 of *Ganja* per puff at an interval of 3 to 4 hours. An average 6 puffs amounting to about 300 mg of delta^9-tetrahydrocannabinol (THC) were consumed daily for a period from 2 to 25 years. Bradycardia and hypotension were consistently found not just after intake but as a persistent effect of the drug in chronic users. Fasting blood sugar was significantly low in cannabis users as compared to controls. Enhancement of salicylate metabolism was also observed in these subjects. (Singh N *et al., Quart J Crude Drug Res*, 1981, 19: 81-91. *Cf. Reviews on Indian Medicinal Plants*, ICMR, Vol. 5: 321.)

Effect on Spermatogenesis

Marijuana decreases spermatogenesis, sperm motility, and increases abnormal forms of sperms. (*PDR for Herbal Medicines*, 2007: 564.)

Toxicity of Maternal Use of Marijuana

In a case-control study, data suggested that maternal use of marijuana may play in etiological use in childhood acute nonlymphoblastic leukemia and may be specific for morphologically defined subgroups. Exposed cases were younger at diagnosis of acute nonlymphoblastic leukemia, as compared to cases not exposed to marijuana; exposed cases were also more often of the myelomonocytic and monicytic subtypes.

In an earlier study by the Children's Cancer Society Study Group, data demonstrated that he risk of developing acute nonlymphoblastic leukemia increased 10-fold among children whose mothers use marijuana during pregnancy. (*PDR for Herbal Medicines*, 2007: 865.) (Cannabis drugs are contraindicated during pregnancy. FDA Pregnancy category C.)

Effect during Pregnancy

The principal psychoactive substance in marijuana, D-9-tetrahydrocannabinol (THC), rapidly crosses the placenta and may remain in the body for 30 days before excretion, thus prolonging potential foetal exposure. THC is also secreted in breast milk. Marijuana smoking produces higher levels of carbon monoxide than tobacco, which is hypothesized to be a potential mechanism of action of prenatal marijuana exposure's impact on the developing foetus. (Julian K. Davies *et al.*, Prenatal Alcohol and Drug Exposures in Adoption, *Pediatr Clin N Am,* 2005, 52: 1369–1393.)

Capparis spinosa Linn.
Vyaagranakha, Caper Bush

Renal Toxicity of High Dosage

To evaluate hepatic and renal toxicity of percolated extract of *Capparis spinosa*, the doses of 200, 400 and 800 mg/kg of methanolic extract were administrated by oral gavages for 7 days in rats. Normal saline, 5ml/kg was given to the control group. Each group contained 6 male rats. On the 8th day, serum and urine samples were collected for liver function tests (ALT, AST, ALP) and renal function tests (BUN, Creatinine, urine ALP and ALP.) The livers and kidneys were isolated for histopathological studies. There were no significant differences in ALT and AST levels of the control and extract treated groups, but there was a significant increase in ALP levels only at doses of 200 mg/kg ($p<0.05$.) The histopathological studies of livers showed no evidence of hepatotoxicity at dose of 200 and 400 mg/kg. Renal function tests including BUN (Blood Urea Nitrogen) and Creatinine were significantly increased after oral administration of 400 and 800 mg/kg ($p<0.05$.) The histopathologic studies of kidneys showed evidence of renal toxicity at doses of 400 and 800 mg/kg. The methanolic extract exhibited no liver toxicity, but resulted in renal toxicity, especially in high dosage. (Haidari *et al.*, *Journal of Shahid Sadoughi University of Medical Sciences*, March-April 2010, 18(4): 47-55.)

Teratogenic Effect

In an experiment, 32 pregnant Balb/c mice were randomly divided into four groups including control and experimental groups. The experimental groups received 200, 400, and 800 mg/kg hydro-alcoholic extract doses of the leaves of *Capparis spinosa*. The control group received the extract doses and urban water. The 18-day embryos, removed out of the uterus, were investigated in the physical deformities. Alizarin staining method was used to assess the embryonic skeletal system. Data was analyzed using one-way ANOVA and Newman-Keuls method. There was a significant reduction in the mean weight of the pregnant mice at the 18th day of pregnancy in 800 mg/kg extract group compared to control group ($p<0.05$.) In addition, there were significant reductions in the mean height and weight of the embryos of the mice in 800 mg/kg extract group than the control group ($p<0.05$.) There was no embryonic physical and skeletal deformation in the experimental groups. 800 mg/kg hydro-alcoholic extract of the leaves of *Capparis spinosa* affected height and weight of the embryos of Balb/c mice and might have teratogenic effects on embryo. (Davari SA *et al.*, Teratogenic effects of hydro-alcoholic extract of *Capparis spinosa* leaf on mice, *J Ethnopharmacol*, 2010, 127(2): 457-62. *Quarterly of Horizon of Medical Sciences*, 2016, 22(2): 95-101.)

Capsicum annuum Linn.

Katuvira, Chilli

Effect of Capsaicin on Bioavailability of Drugs

The bioavailabilities of aspirin (acetylsalicylic acid) and of salicylic acid were studied in male Wistar rats after acute and chronic administration of a *Capsicum annuum* extract, containing 100 mg of capsaicin per gram. With a single administration of 100 mg/kg of the extract, aspirin blood levels remained unchanged, but salicylic acid bioavailability was reduced in 44 per cent compared with control animals. With a single administration of 300 mg/kg of the extract, aspirin blood levels were undetectable while salicylic acid bioavailability was reduced in 59 per cent. Chronic administration once daily for 4 weeks of 100 and 300 mg/kg of the extract resulted in undetectable aspirin blood levels, while salicylic acid bioavailability was reduced in 63 and 76 per cent, respectively, compared with controls. Results show that Capsicum ingestion reduces oral drug bioavailability, likely as a result of the gastrointestinal effects of capsaicin. (Cruz L *et al.,* Ingestion of chilli pepper (*Capsicum annuum*) reduces salicylate bioavailability after oral aspirin administration in the rat, *Can J Physiol Pharmacol.* 1999 Jun; 77(6): 441-446.)

Interaction of Capsaicin with Drug-Metabolization

The interaction of capsaicin with microsomal drug-metabolizing systems was assessed to determine the role that bioactivation of capsaicin may play in the induction of hepatotoxicity and neurotoxicity. Capsaicin produced a type I spectral change in rat hepatic microsomes in a high affinity (Ks=8 microM) concentration-dependent manner and was approximately equipotent with SKF-525A in inhibiting ethylmorphine demethylation. Capsaicin (10 mg/kg, s.c.) inhibited biotransformation *in vivo* as measured by prolongation of pentobarbital sleep time. Reactive metabolites of capsaicin were studied using [3H]dihydrocapsaicin. [3H]Dihydrocapsaicin bound irreversibly to hepatic microsomal protein after *in vitro* incubation or *in vivo* administration. No binding was observed in spinal cord or brain. Although the bioactivation and subsequent covalent binding of capsaicin equivalents may initiate events associated with the hepatotoxicity of capsaicin, it appears that capsaicin-induced neuropathy does not involve covalent interactions with neuroproteins in spinal cord or brain. (Miller MS *et al.,* Interaction of capsaicinoids with drug-metabolizing systems, relationship to toxicity, *Biochem Pharmacol,* 1983 Feb 1, 32(3): 547-551.)

Contraindications

If being treated with any of the following medications, before taking cayenne supplements advice of healthcare provide should be taken:

ACE inhibitors: Captopril (Capoten), Elaropril (Vasotec),Fosinopril (Monopril), Lisinopril (Zestril.) Stomach acid reducers: Cimetidine (Tagamet), Esomeprazole (Nexium), Famotidine (Pepcid.) Omeprazole (Prilosec), Ranitidine (Zantac.) Aspirin: Capsaicin may increase the risk of bleeding associated with aspirin. Blood-thinning medications and herbs: Capsaicin may increase the risk of bleeding associated with certain blood-thinning medications such as warfarin (Coumadin), clopidogrel (Plavix), and herbs such as ginkgo, ginger, ginseng, and garlic. Medications for diabetes: Capsaicin lowers blood sugar levels, raising the risk of hypoglycaemia. Theophylline: Regular use of cayenne may cause body to absorb too much theophylline for asthma. This could be dangerous. (*University of Maryland Medicine Center.*)

Carica papaya Linn.

Eranda karkati, Papaya

C. papaya is not a classical Ayurvedic drug. It was introduced into India during the sixteenth century (the period of the last classical treatise of Ayurveda, *Bhavaprakasha*. *Ayurvedic Pharmacopoeia of India*, Part I, Vol. VI, gave it a non-classical pharmacopoeial name, Erand karkati. In Indo-Unani medicine, Papita Desi is the pharmacopoeial name, while Papita Vilayati is equated with *Strychnos ignatii* Berg. (Ignatus bean.)

Antifertility Effects of Papaya Seeds

In pre-clinical acute and sub-chronic toxicity studies of the methanol sub-fraction (MSF) of the seeds of *Carica papaya*, a putative male contraceptive, have been investigated in rats to evaluate safety of the test substance. A single oral dose of MSF at 2000 mg/kg body weight was studied over 14 days for acute toxicity, and daily oral doses of 50, 100, 250 and 500 mg/kg body weight were studied for 28- and 90-day periods for sub-chronic toxicity. Sperm analysis, hematology, serum clinical biochemistry, libido and pathological examination of vital organs were recorded at the termination of the experimental periods. No overt general toxicity was observed in exposed animals. Food and water intake showed daily fluctuations within control limits. Sperm density showed a significant decrease in all 28- and 90-day repeated dose treated animals whereas total sperm motility inhibition was observed at 250 and 500 mg/kg dose levels at the 28-day time interval but in all dose groups at the 90-day interval. (Nirmal K. Lohiya *et al.*, *Reproductive Toxicology*, October 2006, Volume 22(3): , 461-468.)

In a study in mice, administration of 50 mg/kg/day showed 100 per cent sterility after 60 days, and safety was evidenced after 360 days by unaltered health status, organ weight, hematology, and clinical chemistry, and by an increase in body weight. All altered parameters, including per cent fertility, were restored to control levels in 120 days after treatment withdrawal. (Manivannan B. *et al.*, *Asian J Androl*, 2009, 11(5): 583-599.)

Toxicity of Papain

(Papain enzyme is used for inflammation and edema following trauma and surgery, also as a digestive aid. Raw papain is a mixture of the proteolytic enzymes papain, chymopapain A and B, and papaya peptidase A isolated from the fruit. Papaya leaf contains 2 per cent papain and carpain.)

Orally, large amounts of papain can cause esophageal perforation. Concomitant of papain and papaya can increase the effects and adverse effects of papain. People typically use papaya with enzyme chewable tablets which contain 250 mg of papaya powder, 150 mg of pineapple juice powder, and 10 mg papain, three times, preferably after meals. (*Natural Medicines Comprehensive Database*, 2013: 1201, 1202.)

Toxic Effect of Seeds on Mammalian Vascular Smooth Muscle

Papaya seed extract may exert potentially toxic effects on mammalian vascular smooth muscle. Benzyl isothiocyanate (BITC), the chief bioactive ingredient in seeds, irreversibly inhibits the contraction of dog carotid artery. (*The Review of Natural Products*, Wolters Kluwer, 7th Edn, 2012: 1225.)

Cytotoxic Activity of High Concentration of Papaya Extract

Calcium imaging studies using mammalian cultured endothelial cells showed strong influxes of Ca^{+2} into the cells in response to addition of the papaya seed extract. It was concluded that these extracts, when present in high concentration, are cytotoxic by increasing the membrane permeability to Ca^{2+}, and that the vascular effects of papaya seed extracts are consistent with the notion that benzyl isothiocyanate (BITC) is the chief bio-active ingredient. (Wilson RK *et al., Life Sci.* 2002 Jun 21, 71(5): 497-507.)

Carthamus tinctorius Linn.
Kusumbha, Safflower

Effects Safflower extract on Female Reproductive Hormones in Mice

A study investigated the possible effects of *C. tinctorius* extract on the ovarian histomorphology and the levels of female reproductive hormones in the mice. Sixty adult female Balb/C mice were selected and randomly divided into one control and three experimental groups (n=15.) The control group received only distilled water, while experimental groups were administered intraperitoneally *C. tinctorius* extract at doses of 0.7, 1.4, and 2.8 mg/kg/day for 49 consecutive days. In the end of experiments, blood samples were collected and the sera were analyzed for the levels of FSH, LH, estrogen, and progesterone. The findings showed that treatment with different concentrations of *C. tinctorius* extract reduced the number of ovarian follicles but number of atretic follicles showed an increase. The number and size of the corpora lutea were not affected by extract administration. In addition, in the treated mice with *C. tinctorius* extract, the thickness of the tunica albuginea was increased but the relative and absolute weights of the ovaries decreased significantly. The blood levels of the FSH and estrogen were decreased in the three experimental groups compared with those of the control animals. Thus, the findings of this study indicated that treatment with *C. tinctorius* extract has detrimental effects on the ovarian histomorphology and female reproductive hormones. (Ali Louei Monfared and Amir Parviz Salati, *Avicenna J Phytomed*, 2013 Spring, 3(2): 171–177. PMCID: PMC4075700.)

Carum carvi Linn.

Krishna Jeeraka, Caraway

Effect of *Carum carvi* on TSH Level

In a case, for the first time, *Carum carvi* was found to induce hypothyroidism. To reconfirm the effect of the drug, one of the researchers with history of hypothyroidism tried to test it on her. She was a 24-years-old girl (weight=45 kg) with a history of hypothyroidism for 5 years and had been treating with 100 mcg/day of levothyroxine. TSH level was between 2.5-3.7 mIU/l. All clinical signs and symptoms were assessed and none of hypothyroid symptoms were observed. She was not taking any other medication and her mother had a history of hypothyroidism as well. Her initial TSH level before starting *Carum carvi* was 2.3 mIU/L, indicating euthyroidism. She started to take *Carum* with dose of 1800 mg/day (40 mg/kg) divided in 3 doses. The dose of 40 mg/kg was 1 per cent of the maximum safe dose of *Carum carvi* in rats. Levothyroxine was ingested in fasting state in early morning and *Carum* capsules were used after each meal (breakfast, lunch and dinner.) After 2 weeks, TSH level was 26 mIU/l and thyroid hormone levels were decreased indicating hypothyroid state. TSH level increased to 110 mIU/L, and thyroid hormone level was further decreased. Thyroid exam was unchanged while she was in hypothyroid state both biochemically and clinically. *Carum carvi* was discontinued and thyroid values was measured after 2 weeks and showed TSH level of 25 mIU/L. Dose of Levothyroxine was not changed and thyroid values were measured again 6 weeks after discontinuation of *Carum carvi*. The TSH level further decreased to 11 mIU/l and T_4RIA level was increased to 9 ng/L. Two months later, TSH was further decreased to 7 mIU/l with no intervention indicating subclinical hypothyroid state. The patient obtained 7 scores in Naranjo causality algorithm (probable ADR) and categorized as "certain ADR" in WHO-UMC score. (Seyede Maryam Naghibi *et al., Daru Journal of Pharmaceutical Sciences*, 2015, 23(1): 5.)

Celastrus paniculatus Willd.
Jyotishmati, Staff Tree

Antifertility and Antispermatogenic Effect

The seed oil when given in a dose of 0.2 ml/animal/48h to adult albino rats for 30 days showed antispermatogenic effects as evidenced by vacuolization of seminiferous tubules, germ cell depletion and exfoliation culminating into an arrest of spermatogenesis. The shrunken tubules revealed only sertoli cells and spermatogonia in the final stage of impairment of spermatogenesis. A higher accumulation of total lipids, phospholipids and cholesterol was found in testes of treated rats. (Wangoo D and Bidwai PP, *Fitoterapia*, 1988, 59: 377-382.)

An oily extractive from the seeds showed vacuolization, germ cell depletion and arrest of spermatogenesis in the testes and focal neurosis in liver of rats. (Bidwai *et al.*, *J Ehnopharmacol*, 1990, 28: 293-303. *Cf. Reviews on Indian Medicinal Plants*, ICMR, Vol. 5: 924.)

Hepatotoxicity and Nephrotoxicity

The semipolar and polar compounds present in chloroform and methanol extracts of the seeds, respectively at dose of 0.6 ml in 0.6 ml of edible oil/animal/wk *i.p.* for one month were studied for their effects on liver and kidney in rats, 0.20 and 40 d post treatment. The histopathological studies showed that both fractions induced transient fatty degeneration in the liver and transient proximal tubular damage in kidneys. The increase in serum enzymes (SGOPT, SGPT and alkaline phosphatase) activities supported histomorphological observations. The biochemical and histomorphological observations showed that menthol extract treatment does not affect much the liver and kidney functions, whereas chloroform extract although showed damage to both liver and kidney initially, was followed by repair and restoration. (Bidwai, *Fitoterapia*,1990, 61: 417-424. *Cf. Reviews on Indian Medicinal Plants*, ICMR, Vol. 5: 924.)

Effect of Seeds on Cerebellum in Ageing Albino Rats

In a study, 3 month old (young control), 12 month old (early age- control), 20 month old (late age-control) and 20 month old (late age-treated) male Wistar albino rats (weight 120-340g) were used. Ethanolic extract of seed of C. *paniculatus* (2 g/kg/body weight) was orally administered for 16 days in 20 month old albino rats. The results were compared with 3 month, 12 month and 20 month old control rats. The concentration of trace elements was determined by atomic absorption spectrophotometer. Significant variation was observed in the concentration of trace elements.

In case of copper there was decrease in content in early aged (0.240 ± 0.004) control and age control (0.115 ± 0.004) rats whereas an increase in treated aged rats (0.124 ± 0.004) was observed.

Non-significant variation was observed in zinc content. Young control rats possessed 0.683 ± 0.004 (µg/ml) zinc contents in cerebellum. Age control animal showed the highest level of Zn 0.954 ± 0.002. *Celastrus paniculatus* treated rat show revealed the lowest level of zinc 0.457 ± 0.003 (µg/ml) in cerebellum.

Young control rat had 0.066 ± 0 (µg/ml) manganese content which was significantly decreased in early age control (0.022 ± 0.0008) followed the significant increase in age control (0.087 ± 0.002.) Treated rats possessed the decreased content than age control but higher than young and early age control.

Non-significant decrease in cobalt content was observed during ageing as in young control the highest cobalt content was 0.084 ± 0.0007 followed by decrease in early age control 0.83 ± 0 and age control 0.006 ± 0.0007 (µg/ml.) Treated rats showed an increase in cobalt content up to 0.032 ± 0.0007. (Saini K *et al.,* Effect of *C. paniculatus* on trace elements of cerebellum in ageing albino rats, *Ann Neurosci,* 2012 Jan,19(1): 21-24.)

Centella asiatica (L.) Urban

Manduukaparni, Indian pennywort

Effect on Central Nervous System

In a rodent study an extract of *C. asiatica* was given orally to produce protection from convulsions and to prolong pentobarbital sleeping time. The antidepressant effect was thought to be via D-2 receptors and a cholinergic mechanism. (*PDR for Herbal Medicines*, 2007: 401.)

In the tests on oxidative stress parameters only 200 and 300 mg/kg (in male rats) produced significant decreases in brain levels of malondialdehtde (p<0.001) with simultaneous increase in glutathione. Catalase levels increased significantly in group treated with 200 mg/kg (p<0.05) and 300 mg/kg (p<0.001), but no changes were observed in superoxide dismutase levels.

Oral administration of an aqueous extract from *Centella* fresh leaf to rats at 0.18/kg for 15 days significantly improved performance compared to control animals in a two-compartment passive avoidance test with respect to learning. Concentrations of noradrenaline, dopamine, 5-HT and their metabolities in the brain decreased significantly in the *Centella*-treated group (p<0.01 to p<0.001. (*ESCOP Monographs*, Second Edn Suppl, 2009: 39.)

Hepatotoxic Episodes

Pharma companies must rule out hepatotoxic contaminants before releasing *Centella* products. At least three cases of hepatotoxicity have already been reported. A 61-year-old woman developed elevated liver function tests (LFT.) LFTs improved when *Centella* preparation was discontinued. When the patient took the *Centella* product again she developed LFTs after 2 weeks. In other two cases, a 52-year-old woman and a 49-year-old woman developed symptoms of hepatitis. LFTs were normalized after discontinuation of the drug. (*Natural Medicines Comprehensive Database*, 2013: 764.)

Cinnamomum tamala Nees & Eberm.

Tamaalpatra, Indian Cinnamon

Cardioprotective Effect

A study was designed to scientifically evaluate the cardio protective potential of ethanolic extract of dried leaves of *Cinnamomum tamala* (EECT)) doxorubicin induced myocardial infarction in Wistar albino rats. Myocardial infarction was produced in rats with 15 mg/kg of doxorubicin administered intraperitoneally (i.p), in six divided doses for two weeks. Effect of oral treatment of EECT at two doses (200 and 400 mg/kg body weight), both in prophylactically and curatively manner was evaluated against doxorubicin (15 mg/kg, i.p) induced myocardial infarction. Levels of marker enzymes, Creatinine Phospho Kinase (CPK), Lactate Dehydrogenase (LDH), Alanine Amino Transferase (ALT) and Aspartate Amino Transferase (AST) were estimated in both the serum and heart tissues; antioxidant parameters *viz.*, catalase (CAT) and malondialdehyde (MDA) were assayed in heart homogenate. Doxorubicin significantly increases the serum levels of marker enzymes and reduction of endogenous antioxidants when compared with normal rats. EECT elicited a significant cardioprotective activity by lowering the levels of serum marker enzymes and lipid peroxidation and elevated the levels of catalase. (Nagaraju B *et al., International Journal of Biomedical and Advance Research*, 2016.)

Effects on Thyroid Hormones

The effects of betel leaf extract (0.10, 0.40, 0.80 and 2.0 g kg-1 day-1 for 15 days) on the alterations in thyroid hormone concentrations, lipid peroxidation (LPO) and on the activities of superoxide dismutase (SOD) and catalase (CAT) were investigated in male Swiss mice. Administration of betel leaf extract exhibited a dual role, depending on the different doses. While the lowest dose decreased thyroxine (T_4) and increased serum triiodothyronine (T_3) concentrations, reverse effects were observed at two higher doses. Higher doses also increased LPO with a concomitant decrease in SOD and CAT activities. However, with the lowest dose most of these effects were reversed. These findings suggest that betel leaf can be both stimulatory and inhibitory to thyroid function. (Panda S, Kar A., Dual role of betel leaf extract on thyroid function in male mice, *Pharmacol Res.* 1998 Dec, 38(6): 493-6.)

Effect on Prostatic Hyperplasia

In a study, *Cinnamomum tamala, Berberis aristata,* and *Aconitum heterophyllum* were studied for mechanisms of prostatic hyperplasia in rats. Prostatic enlargement was induced in castrated rats by testosterone injection *s.c.* for 21 days and simultaneously plants drugs were administered orally, daily. On day 22, rats were sacrificed and prostate was removed; weight and volume of prostate was measured; histopathology performed.

Cinnamomum tamala showed significant effect. It reduced prostatic enlargement and improved hyperplastic changes. *Berberis aristata* and *Aconitum heterophyllum* did not show any significant effect. All of them showed mild to moderate anti-inflammatory activity. (Rahul K Dumbre, *Pharmacognosy res,* 2014 Apr-Jun, 6(2): 127–132.)

Interaction with Gentamicin

In a study, the reno-protective properties of *Cinnamomum tamala* against gentamicin-induced nephrotoxicity in rabbits was investigated. Rabbits were randomly divided into four groups (n=6) including Group-1 (normal saline), Group-2 (gentamicin, 80 mg/kg/day), Group-3 (*C. tamala*, 200 mg/kg/day) and Group-4 (gentamicin, 80 mg/kg/day and *C. tamala*, 200 mg/kg/day.)

Gentamicin-treated animals showed significant renal damage as indicated by rise in blood urea nitrogen (54.18 ± 2.60 mg/dl), serum creatinine (4.02 ± 0.14 mg/dl), serum uric acid (2.34 ± 0.12 mg/dl), urinary proteins (3.86 ± 0.32 mg/dl) and decrease in creatinine clearance (0.76 ± 0.09ml/min), urinary volume (126.00 ± 9.09 ml) and body weight (10.80 ± 1.09 per cent.) However, gentamicin and *C. tamala* significantly protected rabbit kidney from structural and functional changes associated with gentamicin. (N. Ullah *et al., Tropical Journal of Pharmaceutical Research,* May 2013, 12(2): 215-219.)

Cinnamomum verum Persl.
Tvak, Ceylon Cinnamon

Subacute and Chronic Toxicity

Cinnamaldehyde added to the diet of rats for 16 weeks at 1 per cent resulted in slight swelling of hepatic cells and slight hyperkeratosis of the squamous portion of the stomach; the no-effect level was 0.25 per cent. (*ESCOP Monographs*, Second Edn: 94.)

Cinnamon-statin Interaction

A 73-year-old woman complained of abdominal pain associated with vomiting and diarrhea after she started taking cinnamon supplements for about 1 week. The patient had been taking rosuvastatin 40 mg orally once a day for coronary artery disease for many months. The laboratory workup and imaging studies confirmed the diagnosis of hepatitis. The detail workup did not reveal any specific cause. Cinnamon-statin interaction was held responsible for her ailment. A few weeks after discharge, the statin was resumed without any further complications. This led to a diagnosis of cinnamon-statin combination-induced hepatitis. (Daniel Brancheau *et al.*, Do Cinnamon supplements cause acute hepatitis? *Am J Case Rep*, 2015, 16: 250-254.)

Interaction with Tetracycline

C. cassia bark extract (2 g in 100 ml) markedly decreased the *in vitro* dissolution of tetracycline hydrochloride. In the presence of *C. cassia* bark, only 20 per cent of tetracycline was in solution after 30 minutes, in contrast to 97 per cent when water was used. (*WHO Monographs on selected medicinal plants*,Vol. 1, 1999: 101.)

Spermicidal Activity

Cinnamon bark oil showed spermicidal effect on human spermatozoa *in vivo* with a minimum effective concentration of 1: 400 V/V. (Buch JG *et al., Indian J Med Res*, 1988, 87: 361-363.)

Major components of both *C. verum* and *C. cassia* in cinnamaldehyde. (*WHO Monographs on selected medicinal plants*, Vol 1, 1999: 95.)

Citrullus colocynthis Schrad.

Indravaaruni, Colocynth Bitter Apple

Effect on Liver

Citrullus colocynthis extract (CCT), in higher concentrations, seem to have some degree of hepatotoxicity. Male wistar rats that were fed diets containing 10 per cent CCT ripe fruits showed enterohepatonephrotoxicity.

In the study, the effect of different concentrations of CCT on the liver was investigated. The results showed some histological changes in the nucleus and cytoplasm of hepatocytes. The changes observed in the nuclei included karyorrhexis and chromatolysis. The mechanism for these changes was not clear but other reports had shown that CCT has a damaging effect on different cells. The ethanol extract of CCT decreases the concentration of sialic acid in serum of mice. This decrease is concomitant with an increase in the unmasking of galactose residues that is recognizable by macrophages in apoptotic cell. It seems that CCT by decreasing sialic acid induces cell degeneration. In addition, CCT causes an increase in neutrophils which confirms the above findings. Inflammation can be precursor of liver fibrosis. In this study, scattered neutrophil and lymphocytes in liver parenchyma were observed which could potentially lead to liver cirrhosis. (Farzaneh Dehghani *et al., Iranian J Pharmacol Therapeutics*, 2006, 117-119.)

A Case Report

A 48-year old man with acute *Citrullus colocynthis* toxicity was admitted to the emergency department ten hours after ingestion of decoction of plant fruit. He developed watery diarrhea, hypotension and hypoglycemia. Hepatic injury was also developed indicated as rising in Alanine aminotransferase (ALT) and Aspartate aminotransferase(AST) enzymes. After supportive management, all parameters reached to the point of normalcy after four days. (M. Rezvani *et al., Indian Journal of Forensic Medicine and Toxicology*, July 2011, 5(2): 25-27.)

Colchicum luteum Baker

Hiranyatuttha, Suranjaan Talkh

Indian substitute of *C. autumnale* Linn., Meadow Saffron, was approved by *German Commission E* for gout.

Comparative Study of *Colchicum luteum* and Allopurinol

The aim of this research was study the lowering effect of uric acid by the use of dried powder of *Colchicum luteum* and allopathic drug allopurinol in rabbits. 12 adult rabbits were divided into three groups A, B and C. Group C was taken as control. The herbal medicine, dried corm powder of *Colchicum luteum* 2.5 mg/kg,/day and dried powder of allopurinol 2 mg/kg/day was used in the study. It was found that dried corm of *Colchicum luteum* significantly reduced the uric acid in adult rabbits as reduced by allopathic medicine, allopurinol. In the light of these findings, it was concluded that the herbal medicine can be used in lieu of the allopathic drug. (Mohammad IS *et al.*, *Acta Pol Pharm*, 2014 Sep-Oct, 71(5): 855-859.)

Toxicity of Prolonged Use

Fresh corms of *Colchicum luteum* contain colchicine (0.20-0.40 per cent.) Colchicine analogs, decteyl thiocolchicine (DTC), decetyl methylcolchicine (DMC) and trimethylcolchicinic acid (TMCA) were found effective in the treatment of gout. Prolonged use of *Colchicum luteum* may cause agranulocytosis, aplastic anaemia and peripheral nerve inflammation. Larger doses are toxic. Even 7 mg of colchicine may prove fatal. (Khare CP, *Indian Medicinal Plants: An Illustrated Dictionary*, 2007: 165-166.)

Adverse Reactions of *Colchicum autumale*

Signs poisoning, including stomachaches, diarrhea, nausea, vomiting, and less frequently, stomach and intestinal hemorrhages can occur even with the administration of therapeutic doses.

Kidney and liver damage, hair loss, peripheral nerve inflammation, myopathia, and bone marrow damage with their resulting symptoms (leukopenia, thrombocytopenia, megaloblastic anemia, and more rarely, aplastic anemia) have been observed following long-term administration. (*PDR for Herbal Medicines*, 2007: 215.)

Dosage for Acute Attack of Gout

Initial oral dose corresponding to 1 mg colchicine, followed by 0.5-1.5 mg every 1-2 hours until pain subsides. Total daily dose must not surpass 8 mg of colchicine. (*The Complete German Commission E Monographs*, American Botanical Council, 1998: 86.) For prophylactic and therapeutic purposes, the dosage should correspond to 0.5 to 1.5 mg of colchicine. (*PDR for Herbal Medicines*, 2007: 215.)

Commiphora mukul (Hk. ex Stocks) Engl.

Guggulu, Indian Bdellium

Efffects on Thyroid Function

The structure of a ketosteroid, isolated from a petroleum ether extract of *Commiphora mukul*, (4,17(20)trans pregnandiene 3,16-dione; Z-guggulsterone) has been established from its physico-chemical properties, specially UV, IR, NMR and mass-spectra. The isolated ketosteroid showed a strong thyroid stimulatory action when administered to albino rats. Its administration (1 mg/100 g body weight) for 6 days brought about an increase in iodine-uptake by thyroid (p <0.05) and enhanced activities of thyroid peroxidase and protease (p <0.001) as well as oxygen consumption by isolated slices of liver and biceps muscle. (Tripathi YB, Malhotra OP, Tripathi SN, Thyroid Stimulating Action of Z-Guggulsterone Obtained from *Commiphora mukul*, *Planta Med.* 1984 Feb, 50(1): 78-80.)

Intragastric administration of steroidal extract of 200.0 mg/kg bw of the oleo-resin per day for 15 days to the female albino mice induced the triiodothyonine (T_3) production and increased triiodothyronine: thyroxine ratio. (S. Panda and A. Kar, *Life Sciences*, 1999, vol. 65, 12: 137–141.)

Impairment of Fertility

Intragastric administration of the oleo-gum resin (dose not specified) reduced the weight of rat uterus, ovaries and cervix, with a concomitant increase in their glycogen and sialic acid concentrations, suggesting an antifertility effect. (Amma MK *et al.*, Effect of Gum Gugglu on the reproductive organs of female rats, *Indian Journal of Experimental biology*, 1978. 16: 1021-1023.)

Crataeva nurvala Buch.-Ham.

Varuna, Three-leaved Caper

Effects against Induced Oxidative Damage in Prostate

In a study, the antioxidative potential of *Crataeva nurvala* bark extract against androgen-mediated oxidative stress in male Wistar rats has been studied. Oxidative damage in prostate was induced in rats by the injection of testosterone (100 mg/kg bw) for 3 days followed by injection of chemical carcinogen N-Methyl N-Nitroso Urea (50 mg/kg bw) for 1 week. The oxidative damage in prostate-induced rats were treated with the ethanolic extract of *Crataeva nurvala* bark (150 mg/kg bw) and testosterone injection (2 mg/kg bw) was also continued through the experimental period of 4 months.

The prostate tissue was dissected out for biochemical analysis of lipid peroxidation and enzymic-antioxidants *viz.*, catalase, superoxide dismutase, glutathione peroxidase, glutathione-S-transferase, and glutathione reductase; the non-enzymic antioxidants *viz.*, reduced glutathione, and Vitamin C. The results revealed that testosterone administration induced the oxidative stress in rat prostate; however, in drug (150 mg/ kg bw) supplemented groups, a significant protective effect of *Crataeva nurvala* bark against testosterone-induced oxidative injury was recorded. The study reveals that constituents present in *Crataeva nurvala* bark impart protection against androgen-induced oxidative injury in prostate. (PMID: 24082632.)

Crocus sativus Linn.

Kumkuma, Saffron

Adverse Reactions

The lethal dose of dried stigmas of *Crocus sativus* is reported to be 12.0-20.0 g/ day; however, smaller doses may cause vomiting, uterine bleeding, bloody diarrhea, hematuria, bleeding from nose, lips and eyelids, vertigo, numbness and yellowing of skin and mucous membranes. Oral administration of 5.0 g resulted in localized skin haemorrhages, marked thrombocytopenia, and abnormalities of blood clotting in one patient.

Dried stigmas are contraindicated during pregnancy, may induce uterine contractions. Also contraindicated in bleeding disorders. It inhibits platelet aggregation, should be used with caution in patients taking anticoagulant or antiplatelet drugs. (Dried stigmas of *Crocus sativus* are used as an emmenagogue and for the treatment of amenorrhoea.) (*WHO monographs on selected medicinal plants*, Vol. 3; 132.)

Sub-acute Toxicity of Crocin

Following IP administration of crocin (180 mg/kg) once a day for 21 days, increased platelets and creatinine levels were observed. At the same dose, a reduction in weight, food intake, alveolar size (in lung) and also minor myosin light chain atrophy were detected.

Administration of crocin (90 mg/kg) decreased levels of albumin and alkaline phosphatase (ALP) while increased the level of LDL. Significant pathological lesions weren't observed in main organs (heart, liver, spleen kidney and lung) after exposure to 15, 45, 90 and 180 mg/kg doses of crocin. In another study, the liver toxicity of crocin was examined. The findings showed that IP administration of 50, 100 and 200 mg/kg of crocin once a week for four weeks in rats didn't alter serum parameters including ALT, AST, ALP, urea, uric acid and creatinine, malondialdehyde (MDA) and gluthatione (GSH) content in liver. Also, using histo-pathological examination, no significant toxicity was observed. (Hasan Badie Bostan *et al., Iran J Basic Med Sci.* 2017 Feb; 20(2): 110–121.)

Sub-acute Toxicity of Safranal

Oral administration of safranal (0.1, 0.25 and 0.5 ml/kg) markedly decreased important hematological factors including RBC counts, HCT, Hb and platelets. Reduced levels of cholesterol, triglyceride, ALP with parallel increase of serum urea nitrogen exhibited remarkable effect of safranal on biochemical parameters. In Pathological examinations noticeable lesion in different tissues (heart, liver and spleen) was not observed, while safranal could induce histopathological changes in lung and kidney. Also, the evaluation of immunotoxic effect of safranal didn't show any significant toxicity on humoral and cellular immune system of mice after IP exposure to 0.1, 0.5 and 1 ml/kg doses for 21 days. (Hasan Badie Bostan *et al., Iran J Basic Med Sci*, 2017 Feb, 20(2): 110–121.)

Sub-chronic Toxicity of Saffron

In a study, oral administration of saffron (4000 and 5000 mg/kg) markedly decreased counts of RBC and WBC as well as, hemoglobin level. Increased BUN and creatinine levels, indicating kidney dysfunction, were detected in animals. This result was confirmed by histopathological examination. Additionally, the activity of liver enzymes including ALT and AST, increased. (It should be noted that the administrated doses are high. Saffron extract in lower doses exhibited protective effects in different models. For example, administration of saffron aqueous extract (25, 50 and 100 mg/kg/day, IP for 30 days) protected against ethylene glycol induced calcium oxalate (CaOx) nephrolithiasis in rats. Additionally, the aqueous extract of saffron stigmas (20 and 80 mg/kg, IP) markedly decreased methyl methanesulfonate–induced DNA damage in mice organs. (The sub-chronic toxicity of constituents of saffron is not available in research material.) (Hasan Badie Bostan *et al., Iran J Basic Med Sci.* 2017 Feb; 20(2): 110–121.)

Crotalaria juncea Linn.

Shana, Sun Hemp

Hepatotoxicity

The effect of the ethanolic extract of *Crotalaria juncea* Linn. seeds has been assessed on liver, kidney, spleen and adrenals of adult rats. Results revealed that its administration at a dose of 200 mg/kg caused significant alterations. Wet weight of the organs was reduced. Protein and glycogen contents in all the organs were decreased significantly, whereas, the activity of acid and alkaline phosphatase was increased. Histology revealed remarkable disintegration necrosis and degeneration in the liver. Renal tubular cells showed degeneration and exfoliation. Adrenals showed hypertrophy in the region of zona glomerulosa. In the spleen the number of megakaryotic cells and lymphocytes was increased. Administration of the extract therefore not only damaged the liver but other vital organs too were also affected. (Prakash AO *et al.*, Toxicological studies on the ethanolic extract of *Crotalaria juncea* seeds in rats, *Journal of Ethnopharmacology*, March 1995, 45 (3): 167-176.)

Antiplantation Activity

An ethanolic extract (90 per cent) of the seeds exhibited anti-implantation activity in albino rats when administered at a dose of 200 mg/kg for 7 days after coitus. It also showed significant abortifacient activity when given for 3 days after coitus which may be due to its estrogenic nature. (Prakash *et al., Indian Drugs*, 1993, 19: 611.)

Croton tiglium Linn.
Jayapaal, Purging Croton

Toxicity of Croton Oil

Croton oil produces severe symptoms of toxicity when taken internally or applied externally to the skin. The skin irritant and tumour promoting diterpene esters of the tigliane type (phorbol esters) and the toxins, crotin I and crotin II with molecular weight 40 kD and 15 kD respectively, have been isolated from the seeds. It brings about itching, burning and after a time, blisters. If taken internally, it leads to burning in the mouth, vomiting, dizziness, stupor, painful bowel movements, and ultimately to collapse. (*The Wealth of India*, First Supplement series, Vol.2, 2001: 250.)

Effects of *Croton tiglium* phorbol esters

Croton oil contains several phorbols 12,13 diesters, short-chain esters of phorbol and 4-deoxy-4-alpha-phorbol and a phorbol ester, 12-O-tetradecanoylphorbol-13-acetate (TPA.) The phorbol esters have been implicated as inducers of Epstein-Barr virus activity with which there is strong etiological relationship to some human tumours such as nasopharyngeal carcinoma.

Several short-chain synthetic phorbol esters and phorbol 14-deoxy-4-deoxy-alpha-phorbol esters actually interfere with tumour-promoting activity of TPA. At a dose of 60-250 mcg/kg bw, phorbol esters actually show significant inhibitory activity against P388 lukaemia in mice. (Kinghorn, *J Nat Prod*, 1987, 50: 1009.)

Isoguanosine, isolated from the seeds, showed anti-tumour activity against various cell lines both *in vitro* and *in vivo* tests, and is found to be effective in solid tumours and ascetic tumours. *C. tiglium* is one of the constituents of the natural drug CP_2 (with berberine and other protoberberine alkaloids.) (*The Wealth of India*, First Supplement series, Vol.2, 2001: 250; Second Supplement series, Vol. 1, 2006: 250.)

Curculigo orchioides Gaertn.

Taalamuuli, Black Mushali

Effect on Hyperglycemia-Induced Oligospermia

In a study, the aqueous extract of the *Curculigo orchioides* (CO) rhizome was evaluated for its effectiveness against streptozotocin-induced hyperglycemic stress and subsequent sexual dysfunction due to hyperglycemia in male rats. Six groups with eight male rats in each group were used for this study. After 28 days, the body and organ weights of the animals were recorded. Behavioural analysis of rats was undertaken to observe the effect on mount, ejaculation and intromission (latencies and frequencies) and hesitation time. Blood glucose and serum testosterone levels were determined 28 days past treatment with CO at 100 and 200 mg/kg doses. Glibenclamide and sildenafil citrate were used as positive controls. This deleterious effect of sustained hyperglycemia evidenced in the principle parameters *viz.*, male sexual behaviour, sperm count, penile erection index and seminal fructose content. Antioxidant and anabolic activities of the extract under investigation could be a major attribute in preserving the sexual functions in hyperglycemic male rats. (Thakur M *et al.*, Effect of *Curculigo orchioides* on hyperglycemia-induced oligospermia, *Int J Impot Res*, 2012 Jan-Feb, 24(1): 31-37.)

Uterine Stimulant Activity

Flavone glycoside isolated from the plant, has been found to be powerful uterine stimulant in guinea pigs, rats and rabbits. (Dhawan BN *et al.*, Evaluation of some indigenous drugs for stimulant effect on rat uterus, *Indian J Med Res*, 1958, 46: 808. Dhar ML *et al.*, Screening of Indian Plants for biological activity, *Indian J Exp Biol*, 1968, 6: 232.)

Curcurma longa Linn.

Haridraa, Turmeric

Antifertility Effects

The effects of aqueous extract of rhizome of *Curcuma longa* on the seminal parameters of Swiss Albino male mice were investigated in this study. 36 mice were divided into six groups of six mice each with one group serving as control. Distilled water (0.1 ml) was administrated orally to control group while remaining five groups were fed with 0.1 ml (500 mg/kg bw/day) aqueous extract of rhizome of *Curcuma longa* for 10, 20, 30, 40 and 50 days. After completion of respective days, controls along with treated mice were sacrificed and semen from caudal epididymal part was assessed for seminal parameters such as sperm counts, motility of spermatozoa, seminal pH and mortality of the sperms. Sperm counts, sperm motility and seminal pH of cauda epididymis shows highly significant decline during 30 to 50 days treatment (p<0.001) than the control while mortality of spermatozoa increases significantly (p<0.001) than the control group of mice. Such significant alteration suggest that aqueous extract of rhizome of *Curcuma longa* produced adverse effects in seminal parameters and thus cause infertility among the treated groups of mice. (Anita Raj Hembrom, Antifertility Effects of Rhizome of *Curcuma longa* on Seminal Parameters of Swiss Albino Male Mice, *Research of Pharmacy and Technology*, 2015, Vol.8(4): 404-406.)

In another study, the contraceptive effect of the crude extracts of *Curcuma longa* in adult proven fertile male Wistar rats was assessed. Rats were randomly divided into 3 groups of 10 animals each: Group 1: Control (1 mL distilled water per day); Group 2: Aqueous extract (500 mg/kg^{-1}/day^{-1}) and Group 3: Alcoholic extract (500 mg/kg^{-1}/day^{-1}) After 55 days of treatment, fertility test was performed and 24 h after the last dose at day 60, the animals were scarified. The weights of testes and epididymides were significantly lower (P 0.01) in both the extract treated groups than in the controls. The weights of the seminal vesicle and ventral prostate of the treated groups were also decreased, but statistically insignificant. The sperm motility in cauda epididymides was decreased by 60.13 per cent and 63.39 per cent in the aqueous and alcoholic extract groups, respectively. A highly significant (P 0.01) reduction in sperm density in the testes and cauda epididymides was observed in both the extract treated groups as compared with the controls. Fertility test showed zero fertility in both treated groups. No significant change was observed in blood and serum parameters. Two months after cessation of drug treatment, the sperm count and motility were recovered. (Purohit Ashok, Bhagat Meenakshi, Contraceptive effect of *Curcuma longa* (L.) in male albino rat, *Asian J Androl,* 2004 Mar, 6: 71-74.)

Other Adverse Effects

Turmeric may increase the risk of bleeding or potentiate the effects of warfarin therapy. (Heck AM *et al.*, Potential interactions between alternative therapies and warfarin, *Am J Health Syst Pharm*, 2000, 57(13): 1221-1227.)

Turmeric should be used with caution for gall bladder problems. (Blumenthal *et al.*, *Herbal Medicine: Expanded Commission E Monographs*, 2000, 379–384.)

Turmeric is contraindicated during pregnancy as it can cause uterine stimulation. (*PDR for Herbal Medicines*, 2007: 865.)

Curcumin may Increase Serotonin

The active constituent of turmeric, curcumin is reported in some studies to produce a marked increase in serotonin and noradrenalin levels at 10 mg/kg dose in frontal cortex and hippocampus also increased the dopamine levels in frontal cortex and striatum parts of mice brain. Curcumin was found to inhibit monoamine oxidase activity in the mouse brain. These findings suggest that the anti-depressant effects of curcumin may involve the central monoaminergic neurotransmitter systems. (Xu Y *et al.*, The effects of curcumin on depressive-like behaviours in mice, *Eur J Pharmacol*, 2005; 518 (1): 40-46.)

Comparative Study of Rhizomes and Leaves

In a study, the rhizomes and leaves of turmeric were extracted separately with ethanol by Soxhlet extraction. The percentage yield of rhizomes and leaves of turmeric was 12 per cent and 17 per cent respectively. The study was focused on the isolation of curcuminoids. From TLC, the resolution of Rf value was observed in rhizomes at 0.8, 0.66, 0.51 as Curcumin, Demethoxycurcumin, Bisdemethoxycurcumin respectively, whereas Rf value of leaves was 0.42 as Bisdemethoxycurcumin, when visualized under 366nm under bright yellow fluorescent. The phytochemical screening of leaf extracts showed the presence of flavonoids, cardiac glycosides and phenols. The ash value of turmeric rhizomes and leaves was 3.33 per cent and 6.67 per cent, acid insoluble value was 1.3 per cent and 2 per cent and water insoluble value was 13.3 per cent and 16.67 per cent respectively. The moisture content of rhizomes and leaves in IR- Moisture balance was found to be 0.93 and 0.30 whereas in Tray Drier the moisture balance was 0.46 and 0.27, respectively. The melting point was observed as 160-163°C and 116-120°C respectively. (Seema, Parminderjit Kaur, Comparative study of pharmacognostical and preliminary phytochemical investigation of *Curcuma longa* leaves and rhizomes, *PharmaTutor*, 2016, 4(10): 31-36.)

Bioavailability Issue of Curcumin

A single dose of curcumin administered orally to 8 human volunteers led to undetectable or very low serum concentrations. Concomitant administration of piperine (an inhibitor of hepatic and intestinal glucuronidation to enhance the serum concentration, extent of absorption and bioavailability of curcumin) as a single 20 mg dose significantly increased serum concentration at 0.25 and 0.5 hour.

After oral administration of curcumin to rats at 1 g/kg/bw, about 75 per cent was excreted in the faeces and only traces in the urine; plasma and bile concentrations of curcumin were negligible. Following intravenous application, curcumin was excreted via bile. (*ESCOP Monographs*, second edn 2003: 112.)

A encapsulated liposomal form has been investigated in animals to overcome the limitation of curcumin bioavailability.

Micro and nano liposome vesicles were prepared using a lipid film hydration method and a sonication method. Phospholipid, cholesterol and curcumin were used to form micro and nano liposomes containing curcumin. The size, structure and properties of the liposomes were characterized by using optical microscopy, transmission electron microscopy, and UV–vis and Raman spectroscopy. It was found that the size of the liposomes was dependent on their composition and the preparation method. The hydration method created micro multilamellars, whereas nano unilamellars were formed using the sonication method. By adding cholesterol, the vesicles of the liposome could be stabilized and stored at 4 °C for up to 9 months. The liposome vesicles containing curcumin with good biocompatibility and biodegradability could be used for drug delivery applications. (Tuan Anh Nguyen *et al.*, Micro and nano liposome vesicles containing curcumin for a drug delivery system, *Advances in Natural Sciences: Nanoscience and Nanotechnology*, 5 July, 2016, Volume 7, Number 3. Vietnam Academy of Science and Technology.)

Curcuma caesia Roxb.

Nishaa, Kaali Hadi, Black Turmeric

Curcuma longa's synonyms Rajani, Nishaa, Nishi, Raatri, Nilkanth should be equated with *Curcuma caesia* Roxb. (C. P. Khare.)

Curcuma caesia is ethno-medicinally important herb in South East Asia. The herb is available throughout north-east, central India, Papi Hills of East Godavari, West Godavari, and Andhra Pradesh.

Phytochemical screening of *n*-hexane, petroleum ether (60: 80), benzene, chloroform, ethyl acetate, methanol, and water extract of rhizome of *Curcuma caesia* revealed the presence of alkaloids, phenols, phytosterols, terpenoids, carbohydrates, tannins, glycosides, saponins, quinones, amino acids, oils and flavonoids.

About 30 volatile oil components were identified in the rhizomes of *Curcuma caesia* by GC-MS, representing 97.48 per cent of the oil, with camphor (28.3 per cent), ar-tumerone (12.3 per cent), (Z)-ocimine (8.2 per cent), 1-ar-curcumene(6.8 per cent), 1, 8-cineole (5.3 per cent), element (4.8 per cent), borneol (4.4 per cent), bornyl acetate (3.3 per cent) and curcumene (2.82 per cent) as the major constituents. Rastogi *et al.*, reported linalool as the major component comprising 20.42 per cent followed by ocimine (15.66 per cent), 1- ar-curcumene (14.84 per cent), zingiberol (12.60), 1, 8-cineole (9.06 per cent), and borneol (7.4 per cent) as major constituent. (Rastogi RP, Malhotra BN. *Compendium of Indian medicinal plant,* CDRI: New Delhi, 1999: 241.)

Nishaamalaki Churna of Ashtangahridya (seventh century) was prescribed for diabetes. The powder contained Nishaa and Aamalaki (Emblic myrobalan) in equal quantity. This is the only two-plant drug medicine for diabetes in Ayurveda. (C. P. Khare.)

Antidiabetic Activity of Nishaa

The methanol extract of *Curcuma caesia* rhizome showed antidiabetic activity by lowering blood glucose in *in vivo* studies and controlled intestinal absorption of monosaccharides by inhibiting alpha amylase and alpha glucosidase. It also enhanced the glucose uptake in yeast cells, proving proper glucose utilization. The IC_{50} values of methanol extract of *Curcuma caesia* (MECC) were found to be 442.92±10.05 µg/ml and 95.40±9.74 µg/ml in alpha amylase and alpha glucosidase inhibition respectively. The extract has effective antioxidant activities, by successfully scavenging free radicals like superoxide and hydroxyl ions. The histopathological studies of pancreas showed gradual healing after treatment. While performing the toxicity study no deaths were observed when the animals were given a dose of 2000 mg/kg b.w. (Poulami Majumder *et al.*, Preclinical evaluation of Kali Haldi (*Curcuma caesia*): a promising herb to treat type-2 diabetes, *Oriental Pharmacy and Experimental Medicine*, June 2017, 17(2): 161.) (https://www.researchgate.net/publication/316142190. Accessed Jul 02, 2018.)

Datura metel Linn.

Dattuura, Thorn Apple

Toxic Effects during Pregnancy

In an experiment a single dose of 50 mg/kg/bw of aqueous extract of the seeds administered orally to pregnant rats at different days of pregnancy resulted in significant decrease in the value of RBC, WBC, Hb percentage, fasting blood sugar, acid and alkaline phosphate, proteins and level of cholesterol. The concentration of serum Na^+ and K^+ and the activity of SGOT remained unchanged, whereas that of SGPT increased. The activity of SGPT remained high even after 20d treatment. The alterations were more apparent after 2-3d of administration. And later the values resumed to normal. (*Cf Reviews on Indian Medicinal Plants*, ICMR, Vol. 9: 118-119.)

Toxicity of Overdoses

The intake of very high dosages of *D. stramonium* leads to central excitation (restlessness, compulsive speech, hallucinations, delirium, manic episodes) followed by exhaustion and sleep.

The 4 early warning symptoms of poisoning are skin redness, dryness of the mouth, tachycardic arrhythmia, and mydriasis. Accommodation disorders, heat build-up through decline in sweat secretion, micturition disorders, and severe constipation can occur as side disorders.

Lethal dosage for adults: starting at 100 mg atropine, 15 to 100 g of the leaf drug, 15 to 25 g of the seed drug. (*PDR for Herbal Medicines*, 2007: 483.)

Interaction with Succinylcholine

Concurrent use may result in increased neuromuscular blockade. (*PDR for Herbal Medicines*, 2007: 483.)

Dioscorea bulbifera Linn.

Vaaraahi, Air Yam

Hepatotoxicity of Diosbulbin

In China, a study was undertaken to investigate the hepatotoxicity induced by *Dioscorea bulbifera* in mice. Through the acute toxicity of various extracts including the EtOAc fraction (EF) and the non-EtOAc fraction (Non-EF) from ethanol, and the ethanol itself, it was found that the EF contains the toxic ingredients of *D. bulbifera* rhizome. On this basis, to study the hepatotoxicity induced by the toxic ingredients, mice were treated with 0.5 per cent sodium carboxymethyl cellulose (CMC-Na) alone or the EF of *D. bulbifera* rhizome at doses of 80, 160, 320, and 480 mg/kg once daily i.g. for fourteen consecutive administrations. Serum samples were collected for determination of the biomarkers for liver injury, such as, alanine transaminase (ALT) and aspartate transanimase (AST.) Hepatic tissues were used to assay for the level of lipid peroxide (LPO), amounts of antioxidants such as glutathione, and activities of antioxidant-related enzymes for liver oxidative-antioxidative status in mice. The results showed that ALT and AST were significantly elevated after fourteen consecutive administrations of the EF of *D. bulbifera* rhizome. In addition, the level of LPO increased remarkably, while the glutathione amounts, and the activities of the antioxidant-related and glutathione-related enzymes including superoxide dismutase (SOD), glutathione peroxidase (GPx), glutathione-S-transferase (GST), glutathione reductase (GR) and glutamate-cysteine ligase (GCL) of hepatic tissues all decreased conspicuously, in livers of mice treated with the EF of *D. bulbifera* rhizome. Results indicated that the EF contains the main toxic ingredients of *D. bulbifera* rhizome, and the mechanism of hepatotoxicity induced by it may be due to liver oxidative stress injury in mice. (Wang J *et al., Biosci Trends,* 2010 Apr, 4(2): 79-85.)

Embelia ribes Burm. f.

Vidanga, Embelia

Toxicity of Embelin

Administration of embelin in female cyclic rats at a dose of 120 mg/kg body weight is reported to cause no change in the weight of liver, kidney and spleen, however, the net weight of the adrenals showed a remarkable increase. Administration of embelin for 6 weeks caused severe pathological changes in the liver and kidney which mainly included disintegration, necrotic changes and perinuclear vacuolation. Marked tubular damage was observed in the kidneys. The adrenals showed hypertrophy and the histological features of the spleen remained unchanged. (Anand O. Prakash, *Phytother Res*, 1994, 8: 257.)

Antispermatogenic Effect

Daily subcutaneous administration of the embelin at a dose of 20 mg/kg body weight to male albino rats for 15 or 30 days revealed an inhibition of: , epididymal motile sperm count, fertility parameters such as pregnancy attainment and litter size, and the activities of the enzymes of glycolysis and energy metabolism. These changes were reversible, as seen after 15 and 30 days of recovery. Addition of embelin to epididymal sperm suspensions caused a dose- and duration-dependent inhibition of spermatozoal motility and the activities of the enzymes of carbohydrate metabolism. Both *in vivo* and *in vitro* treatment with the drug causes profound morphological changes in spermatozoa such as decapitation of the spermatozoal head, discontinuity of the outer membranous sheath in the mid-piece and the tail region, and alteration in the shape of the cytoplasmic droplet in the tail. (Gupta S *et al.*, *Contraception*, 1989, 39(3): 307-320.)

Euphorbia neriifolia Linn.

Snuhi, Dog's Tongue

Toxicity of latex

The latex, in distilled water, produced persistant rise in blood pressure of dogs in a dose of 0.1 ml/kg *i.v.* In higher doses (0.2 ml/kg) there was no change or a fall in blood pressure after initial dose. It produced many-fold increase in the rate of respiration in dogs both on intravenous and intracerebroventricular administration. Still higher doses produced apnoea. (Iqbal NJ, *Indian J Pharmacol*, 1969, 1: 8-9.)

Accidental inoculation of latex of *Euphorbia* (species was not known) to three patients exhibited ocular toxicity. The initial symptoms in all cases were severe burning sensation with blurring of vision. Visual acuity reduced from 20/60 to counting fingers. Clinical findings varied from kerato-conjunctivitis, mild to severe corneal edema, epithelial defects, anterior uveitis and secondary elevated intraocular pressure. (Basak SK *et al.*, Keratouveitis caused by Euphorbia sap, *Indian J Opthalmol*, 2009, 57: 311-313.)

Ferula asafoetida Regel.

Hingu, Asafoetida

Effects of Gum-Resin on Biochemical and Histological Parameters

Asafoetida did not show any acute toxicity, but chronic administration could have undesirable effects on hepatocytes and hematological factors. In a chronic study, male Wister rats were administered with various doses of asafoetida (25, 50, 100, and 200 mg/kg body weight) for a period of 6 weeks. At the end of experiment, the effects of asafoetida on haematological, renal, and hepatic markers and histological parameters were analyzed. In acute toxicity study (conducted by the oral administration of 250, 500, and 1,000 mg/kg body weight of the animal), no mortality was seen up to 72 h of the administration of asafoetida. No signs of neurological and behavioural changes were noticed within 24 h. In the chronic study, the asafoetida intake has changed the haematological parameters such as red blood cell (RBC), white blood cell (WBC), hematocrit (HCT), and platelets. Aspartate aminotransferase (AST) and lactate dehydrogenase (LDH) were significantly increased in treated animals. The plasma level of urea and creatinine were not altered by the administration of asafetida throughout the study. Histopathology study indicates hepatotoxicity, but no signs of prominent pathological changes in kidney. (Seyyed Majid Bagheri *et al.,* Evaluation of Toxicity and Effects of Asafetida on Biochemical, Hematological, and Histological Parameters in Male Wistar Rats, *Toxicology International,* 2015 Jan, 22(1): 61.)

Effect on Human Gastric Mucosa

The effect of intragastric infusion of black pepper powder in doses of 0.2, 0.4 g/h and asafoetida in doses of 0.1, 0.2 and 0.4 g/h on different days in human volunteers was studied by measuring the DNA contents of gastric aspirates. The mean increase in the DNA content of gastric aspirates as compared to the basal value for black pepper at 0.2 g/h was 1.00; at 0.4 g/h was 6.30; in the case of asafoetida, the values at 0.1 g/h were 1.18, at 0.2 g/h 4.15 and 0.4 g/h 9.13. The results showed that black pepper does not damage the human gastric mucosa, whereas asafoetida does, as assessed by its effect on the rate of exfoliation of the surface epithelial cells of human gastric mucosa. (Desai and Kalro, 1985, Downloaded from *https: //www.nal.usda.gov/.*)

Nutritional Value Doubtful

The influence of oleo-gum resin of asafoetida was studied on quality and utilization of protein of sorghum and chickpea. Addition of asafoetida did not affect the protein digestibility of sorghum but there was an increase in biological value on addition of asafoetida. The net protein utilization followed the same trend as biological value. The protein digestibility, biological value and net protein utilization of chickpea diet did not show any significant difference in control and asafoetida added diet. (*Reviews on Indian Medicinal Plants*, ICMR, Vol. 11, 2013: 107.)

Antifertility Activity

Methanol extract of the resin, administered orally to Sprague–Dawley rats at a dose of 400 mg/kg daily for 10 days, prevented pregnancy in 80 per cent of the rats. When administered as a polyvinylpyrrolidone 1: 2 complex, 100 per cent pregnancy inhibition was observed at this dose. Lower doses of the extract produced a marked reduction in the mean number of implantations. Significant activity was observed in the hexane and chloroform eluents of sulfur-containing extract in an immature rat bioassay, the methanol extract was devoid of any estrogenic activity. (Keshr G *et al.*, Post-coital antifertility activity of *Ferula asafoetida* extract in female rats, *Pharmac Biol*,1999, 37: 273-278.)

Foeniculum vulgare Mill.
Mishreya, Fennel

Hepatotoxicity

In rats, hepatotoxicity was observed when dietary intake of *trans*-anethole exceeded 30.0 mg/kg bw per day. In female rats, chronic hepatotoxicity and low incidence of liver tumours were reported with a dietary intake of *trans*-anethole 550.0 mg/kg bw per day, a dose about 100 times higher than the normal human intake. In chronic feeding studies, administration of *trans*-anethole, 0.25 per cent, 0.5 per cent or 1 per cent in the diet, for 117-121 weeks produced a slight increase in hepatic lesions in the treated groups. (*Food and Chemical Toxicology*, 1999, 37: 789-811; *ibid*, 1989, 27: 11-20.)

Estrogenic Activity

Oral administration of the acetone extract of the fruits (150 and 250 mcg/100 g for 10 days to female rats led to vaginal cornification and estrus cycle. While moderate doses caused increase of weight of mammary glands, higher doses increased the weight of oviduct, endometrium, myometrium, cervix and vagina also. (*Reviews on Indian Medicinal Plants*, ICMR, New Delhi, Vol. 11: 468.)

Effect on Serum Estradiol

Four case reports suggest that fennel tea given to infants for prolonged periods of time resulted in premature breast development in girls. All the 4 subjects had serum estradiol levels 15 to 20 times higher than normal values for the ages. After stopping the ingestion the premature breast development resolved within 3 to 6 months. (*J Pediatr Surg*, 2008, 43(11): 2109-2111.)

Reproductive Toxicity

Trans-anethole administered to adult female rats on days 1-10 of pregnancy at 50, 70 and 80 mg/kg body weight inhibited implantation by 33 per cent, 66 per cent and 100 per cent respectively. Further experiments at the 80 mg/kg dose level showed that the rats given trans-anethole only on day 1-2 of pregnancy, normal implantation and delivery occurred. No gross malformations of offspring were observed in any of the groups. (*ESCOP monographs*, Second edition, 2003: 165.)

Gloriosa superba Linn.
Laangali, Glory Lilly

A Case Report of Tuber Toxicity

A female aged 35 years was brought to hospital with the complaints of vomiting, numbness of the mouth, burning sensation and dryness of the throat. On admission, her vitals were found to be stable and her biochemical investigations were within normal limits. Within 2 hours of admission, she developed symptoms of bloody diarrhea and dehydration. Her blood pressure was 100/70 mm Hg, pulse rate was 112/minute and the respiratory rate was 20/minute. Gastric lavage followed by symptomatic and supportive treatment was given. After 24 hours of admission, serum urea and creatinine levels were found to be markedly elevated (urea: 70 mg/dl, creatinine: 3.5 mg/dl) and her gastrointestinal symptoms worsened. The patient ultimately died due to shock and multi organ failure after 30 hours. The clinical cause of death was attributed as due to multiple organ failure and acute respiratory distress syndrome.

History revealed that she had consumed 3 to 5 tubers of an unknown plant which was identified to be *Gloriosa superba*. When an autopsy was conducted, internal examination revealed inflammation and multiple hemorrhagic petechial spots on the wall of the stomach and intestines. Liver and spleen were enlarged and congested, small intestine and colon were distended and multiple petechial hemorrhages were also observed throughout the fatty and subserosal tissues. Also numerous discrete petechial hemorrhages were found over the surface of the lungs, kidneys and base of the heart.

(Peranantham S *et al., Fatal Gloriosa superba poisoning – a case report.* https://www.researchgate.net/publication/264348411, accessed Feb 23, 2018.)

(In Ayurvedic medicine, for detoxification, the root is soaked in cow's urine for 24 hours, then washed and dried. Unprocessed tuber contains 0.66-0.92 per cent colchicine, flower 1.05-1.18 per cent, and leaf 0.87-2.36 per cent.)

Glycyrrhiza glabra Linn.
Yashtimadhu, Licorice

Lower Bio-availability of the Root Extract

Plasma levels of glycyrrhetic acid after oral administration of an aqueous root extract to healthy volunteers were found to be lower than those observed after oral administration of a corresponding amount of glycyrrhizic acid. The peak serum concentration of glycyrrhizic acid occurred less 4 hours after administration of the root decoction, and was not detectable after 96 hours. In contrast, the serum concentration of glycyrrhetinic acid reached a maximum after 24 hours and remained detectable in the urine for approximately 130 hours. The low urinary concentration of both compounds suggested excretion via gastrointestinal route. (*ESCOP monographs*, Second edition, 2003: 301.)

Adverse Effects Due to Intoxication of 50 g/day *G. glabra*

Consumption of as little as 50 g/day of *G. glabra* can produce mineralocorticoid hypertension. Mineralocorticoid hypokalemic effects are responsible for a number of case reports of hypokalemic paralysis, pseduoaldosteronism and cardiac myopathy. Several authors are of the opinion that symptoms including severe hypokalemia, mineralocorticoid hypertension, cardiac arrhythmias, paralysis of the extremities, metabolic alkalosis, hypoxemia and hypercapnia occurred due to liquorice intoxication.

The intake of up to 217 mg of glycyrrhizic acid per day (as licorice extract) led to neither clinical effects nor changes in laboratory parameters, suggesting the absence of mineralocorticoid-like activity. Only the highest doses of the extract led to adverse effects in healthy subjects, in particular those with subclinical diseases or in situations favouring sodium retention, such as in the premenstrual period or when taking oral contraceptives. Glycyrrhetic acid administered orally to 10 healthy, normotensive volunteers for 8 days at 500 mg/day, in 2 divided doses, exerted mineralocorticoid activity as shown by significant increases in plasma sodium ($p < 0.01$) and urinary potassium, significant decreases in plasma potassium ($p < 0,01$) and aldosterone ($p < 0.05$.) Urinary excretion of free hydrocortisone was elevated and plasma hydrocortisone levels virtually unchanged in the presence of markedly decreased levels of both plasma cortisone and urinary free cortisone. This provides direct clinical support for hypothesis that glycyrrhetinic acid induces inhibition of the activity of 11beta-hydroxysteroid dehydrogenase, resulting in a blockade in the conversion of hydrocortisone (cortisol) to cortisone. (*ESCOP monographs*, Second edition, 2003: 300.)

Effects on Hormonal Levels

In a latest study, the effect of licorice was investigated on androgen metabolism in nine healthy women 22-26 years old, in the luteal phase of the cycle. They were given 3.5 g of a commercial preparation of licorice (containing 7.6 per cent W.W. of glycyrrhizic acid) daily for two cycles. They were not on any other treatment. Plasma renin activity, serum adrenal and gonadal androgens, aldosterone, and cortisol were measured by radioimmunoassay. Total serum testosterone decreased from 27.8+/-8.2 to 19.0+/-9.4 in the first month and to 17.5+/-6.4 ng/dL in the second month of therapy (p<0.05.) It returned to pre-treatment levels after discontinuation. Androstenedione, 17OH-progesterone, and LH levels did not change significantly during treatment. Plasma renin activity and aldosterone were depressed during therapy, while blood pressure and cortisol remained unchanged. The study indicates that Licorice could be considered an adjuvant therapy of hirsutism and polycystic ovary syndrome. (Armanini D *et al.*, Licorice reduces serum testosterone in healthy women, *Steroids*, 2004 Oct-Nov, 69(11-12): 763-6.)

Comments by Decio Armanini, Guglielmo Bonanni, University of Padua, Italy, and Mario Palermo, University of Sassari, Italy (*Exp Clin Endocrinol Diabetes*, 2003 Sep, 111(6): 341-3): "We have previously found that licorice can reduce serum testosterone in healthy men. These results were not confirmed in another study, where the same amounts of licorice did not decrease salivary testosterone values. In the actual study we treated more cases with the same amount of licorice and reproduced our previous data. The mean testosterone values decreased by 26 per cent after one week of treatment (p < 0.01.) There was also a significant increase in 17-OHP and LH concentrations and a slight, but not significant decrease in free testosterone. Licorice treatment, in addition, did not affect the response of testosterone and 17-OHP to stimulation with beta-HCG."

In October 1999 (*N Engl J Med*, 341: 1158), Armanini D *et al.*, evaluated the effect of licorice on gonadal function in seven normal men, 22 to 24 years of age. The men were given 7 g daily of a commercial preparation of licorice in the form of tablets (*Saila*, Bologna, Italy) containing 0.5 g of glycyrrhizic acid, as determined by gas chromatography–mass spectrometry; the effect on the metabolism of mineralocorticoids in these men was reported previously. Serum testosterone, androstenedione, and 17-hydroxyprogesterone were measured by radioimmunoassay before and after four and seven days of administration of licorice and four days after it was discontinued. During the period of licorice administration, the men's serum testosterone concentrations decreased and their serum 17-hydroxyprogesterone concentrations increased. These results demonstrate that licorice inhibits both 17beta-hydroxysteroid dehydrogenase and 17, 20-lyase, which catalyzes the conversion of 17-hydroxyprogesterone to androstenedione.

A Case of Severe Toxicity

A sixty-one year old man who was admitted to hospital because of severe hypokalemia, rhabdomyolysis and high blood pressure. Severe hypokalemia may lead to rhabdomyolysis. A diagnosis of excess amount of apparent mineralocorticoid was attributable to licorice and grapefruit juice ingestion. Glycyrrhizic acid and glycyrrhetinic acid, its hydrolytic product in licorice extracts, and polyphenols, in grapefruit juice, can inhibit 11 betahydroxysteroid dehydrogenase type 2, the enzyme that converts cortisol to cortisone. (Sardi A *et al., Ann Ital Med Int*, 2002, 17(2): 126-129.)

Shorter Gestation Period Due to Heavy Licorice Consumption

Heavy licorice (glycyrrhizin) consumption has been associated with shorter gestation. Heavy glycyrrhizin exposure was associated with preterm delivery and may be a novel marker of this condition. (Strandberg TE, Andersson S, Jarvenpaa AL and McKeigue PM, Preterm Birth and Licorice Consumption during Pregnancy, *Am J Epidemiol* 2002; 156(9): 803-805.) (*ESCOP Monographs*, Second Edn. Supplement, 2009: 297.)

Suppression of Plasma Renin Activity and Aldosterone Levels

The mechanism by which the glycyrrhizinates exert their effect on renin-angiotensin-aldosterone system has been elucidated. Competitive (and reversible) inhibition of the enzyme 11-beta-hydroxysteroid dehydrogenase results in the suppression of cortisol conversion to inactive cortisone. Consequent suppression of plasma renin activity and aldosterone levels is evident. Exchangeable sodium levels increase and cortisol occupation of mineralocorticoid receptors in the distal kidney tubules is enhanced. The condition responds to administration of spironolactone, potassium supplementation, and discontinuation of licorice. (*The Review of Natural Products*, Wolters Kluwer, 7th Edition, 2012: 993.)

Gmelina arborea Roxb.

Gambhaari, Candahar Tree

Hepato- and nephro-toxicity

A subchronic toxicity study was carried out using three doses of aqueous extract of leaves 31.25 mg/kg, 125 mg/kg and 500 mg/kg of body weight, administrated to rats (5 male and 5 female) for 28-days. At the end of the study, there was neither mortality, nor physiological behaviour changes. The body weight and diet consumption of the rats were not modified. As for blood parameters, a dose dependent increase in cholesterol and transaminases levels and decrease in blood glucose were observed. There was also an increased level of creatinine in male rats and urea in female rats. Histology of the treated rats' liver shows in the centrolobular area lipid inclusions in hepatocytes almost at 500 mg/kg. In kidneys, a thickening of the glomerular interstitium was observed. These results show that the aqueous extract of *Gmelina arborea* is probably hepato- and nephro-toxic. (Osseni R *et al.*, *International Journal of Toxicological and Pharmacological Research*, 2015; 7(2): 116-122.*Cf* ResearchGate.)

Genotoxic Activity

In a study, aqueous extract of *G. arborea* (AEGA) leaves was tested in Swiss albino mice at the dose of 286 and 667 mg/kg bw. Cyclophosphamide 25 mg/kg bw was used as positive control in micronucleus test. The AEGA significantly increased the per cent micronucleated polychrometics at doses of 286 mg/kg and 667mg/kg, after 24, 48, 72h time interval, and also decreased the polychromatic erythrocytes/normochromatic erythrocytes (PCE/NCE) ratio after 24, 48 and 72 h as compared to solvent control group. In this study, the effect of *G. arborea* on mammalian bone marrow cells was investigated using micronuclei formation to assess the genotoxicity of the herb. An increase in the frequency (percentage) of micronucleated polychromatic erythrocytes (MNPCEs) in treated rats is an indication of induced chromosome damages. Decrease in the PCE: NCE ratio due to either direct cytotoxicity or micronuclei formation and heavy DNA damages leading to cell death or apoptosis. AEGA at the tested dose 667 mg/kg and 286 mg/kg bw induced a very significant increase in the frequency of micronuclei and decrease in PCE/NCE ratio after acute treatment, confirming the mutagenic potential of the AEGA extract. (Rohit Sahu *et al.*, *J Adv Pharm Technol Res*, 2010 Jan-Mar, 1(1): 22-29.)

Gossypium herbaceum Linn.
Kaarpaasa, Asiatic Cotton

Effect of Cotton Seed Oil on Pituitary and Testicular Function

In a study, luteinizing hormone releasing hormone (LHRH) 100 micrograms *i.v.*- and Human chorionic gonadotropin (hCG) 3000 IU *i.m.*-stimulation tests were conducted in four controls and in 45 men who had used crude cotton seed oil as their cooking oil. The patients were divided into two groups: group A—17 men with normospermia or oligospermia and group B—28 men who were azoospermic. The basal serum LH and FSH concentrations were within the normal range in group A, whereas those in group B were increased markedly. There was no significant difference in testosterone levels between the two groups, although the levels were significantly lower than in the controls. The response of LH and FSH to LHRH, and of testosterone to hCG stimulation, were within the normal range in group A, whereas in group B the response to the LHRH test was increased significantly while their response to the hCG test was reduced markedly. It was concluded that the functions of the pituitary and Leydig cells remained unchanged in group A after long-term use of crude cotton seed oil, and that once azoospermia has occurred, it is followed by total testicular failure as indicated by the responses to LHRH and hCG tests. (Gui-Yuan Z *et al.*, The effect of long-term treatment with crude cotton seed oil on pituitary and testicular function in men, *Int J Androl*, 1989 Dec,12(6): 404-410.)

Effect of Cotton Seed Extract on Female Rat Uterus

A study was designed to investigate the effects of ethanolic extract of cotton seed on the metabolic activity of uterus and serum level of reproductive hormones in adult female wistar rats. A total of 18 adult female rats weighing between 180 g- 200 g were randomly selected into 3 groups n=6. Group A received phosphate buffered saline, Group B and C were treated with doses of 20 mg/kg and 60 mg/kg body weight of the extract intraperitoneally twice daily at 0700 hours and 1900 hours for 21 days respectively. Animals were sacrificed by cervical dislocation 12 hours after the last administration and blood sample was collected from descending thoracic aorta for estimation of serum levels of estrogen, progesterone, LH and FSH. Uterus excised following abdominal incision, homogenized in 5 per cent sucrose solution to estimate the enzyme activity of Glucose 6- phosphate dehydrogenase and Lactate dehydrogenase.

There was significant decrease ($P < 0.05$) in serum estrogen level; this decrease was more pronounced in the group that was treated with high dosage of 60 mg/kg body weight of the extract. Increase in the serum progesterone level was also observed in the rats treated with dose of 60 mg/kg of the extract when compared with the control rats but significant reduction was observed in the group treated with 20 mg/kg body weight of the extract.

The mean concentration of LH and FSH was significantly lower (P < 0.05) in the rats treated with both 20 mg/kg and 60 mg/kg body weight of the extract when compared with the control rats. Biochemical Analysis revealed significant (P < 0.05) increase in Lactate dehydrogenase activity and increase in G-6-DPH activity of the treated rats when compared with the control rats The activity of carbohydrates metabolic enzymes was statistically higher in the group treated with the higher dosage. The consequent changes observed in the result will interfere with ovulation and implantation thus promoting infertility. (Oyetunji *et al.,* 2012. Anti-fertility effects of cottonseed extract, *World J Life Sci and Medical Research*, 2012, 2(5): 196.)

Gymnema sylvestre R. Br.

Meshashrngi, Australian Cow Plant

Absence of Antidiabetic and Hypolipidemic Effects

In study the antidiabetic and hypolipidemic potential of capsules of dried powdered leaves of *Gymnema sylvestre* (GS) was investigated in Brazil. Wistar rats (*Rattus norvegicus*), weighing 200-250 g were trained to eat a single meal daily at 8: 00-10: 00 (MF rats) during seven days. On day 8, immediately before the meal, the MF rats were divided in two groups, one received *Gymnema sylvestre* (GS) (experimental group), the other received water (control group, i.e, COG group.) After water or GS administration at 8.00 a.m., all rats received a fixed amount of food, *i.e.* 5 g of commercial chow which corresponded approximately 70 per cent of the amount of food ingested by MF rats on the day 8 of feeding training. This procedure and previous feeding training was performed to obtain rats with the ability to ingest the same amount of food at the same period of time (about 30 minutes) like a human meal. Immediately after the meal, the rats were subdivided into 4 subgroups, *i.e.*, rats which were killed by decapitation at 0 (COG: 0 minute and GS: 0 minute) and 30 (COG: 30 minutes and GS: 30 minutes) after the meal. Blood was collected and glucose serum level, triglycerides, cholesterol and total lipids were measured.

In the first set of experiments, the acute effect of (GS) on the elevation of glycemia promoted by a balanced meal in MF rats was investigated. GS (30 mg/kg) did not affect glucose blood level, cholesterol, triglycerides and total lipids, immediately and 30 minutes after a meal with a standard laboratory diet. However, if GS 30 mg/kg was replaced by GS 1000 mg/kg a clear hyperglycemia was observed.

In the second set of experiments, the acute effect of GS on the elevation of glycemia promoted by the administration of 1000 mg/kg amylose (AM) in overnight fasted rats was investigated. 1000 mg/kg GS did not impair the elevation of glycemia promoted by AM. Similar results were obtained if 1000 mg/kg GS was replaced by 500 mg/kg GS or 30 mg/kg GS. In agreement with these results 1000 mg/kg GS did not impair the elevation of glycemia promoted by the administration of 1000 mg/kg glucose.

In the third set of experiments, the acute effect of 30 mg/kg GS administered 30 minutes before, simultaneously or 30 minutes after the administration of soybean oil + 1 per cent cholesterol (SOC) on the elevation of serum lipids in overnight fasted rats was investigated. GS promoted more intense (P<0.05) elevation of triglycerides and total lipids, but not total cholesterol.

In the fourth set of experiments the effect of the treatment during two weeks with 30 mg/kg GS on blood level of glucose and triglycerides, body weight, daily food and water intake were investigated. As shown in non-diabetic rats and alloxan-diabetic rats, these parameters were not affected by the treatment with GS. Similar results were obtained if the treatment with GS was expanded to 4 weeks.

This study highlighted following findings:

GS acutely did not influence the elevation of glycemia promoted by a balanced meal by the administration of amylose or glucose; but promoted more intense ($P<0.05$) elevation of serum lipids after the administration of SOC. Moreover, the sub-acute and chronic treatment with GS in non-diabetic and alloxan-diabetic rats did not change: a) the body weight gain; b) food and water ingestion; c) lipids and glucose blood level. On the other hand, in spite of alloxan-diabetic rats which received GS during 2 weeks showed lower ($P<0.05$) level of TG, this difference is not maintained until 4 weeks of treatment with GS. In addition the level of TG in alloxan-diabetic rats after 2 or 4 weeks of treatment was higher ($P<0.05$) when compared with non-diabetic rats.

Thus, in contrast with several studies which demonstrated antidiabetic (Shanmugasundaram *et al.*, 1990; Kar *et al.*, 1999) and hypolipidemic (Shigematsu *et al.*, 2001; Wang *et al.*, 1998) properties to GS, this study did not find any effect not only after acute administration but also during a sub-acute (2 weeks) and chronic (4 weeks) treatment with GS. The absence of acute, sub-acute and chronic effects could not be attributed to a limitation of the experimental rat model since a clear antidiabetic and antihyperlipidemic effect was routinely obtained in the lab with conventional doses of classical antihyperlipidemic, antihyperglycemic and hypoglycemic (insulin) drugs.

The difference from these results and majority of studies showing antidiabetic and antihyperlipidemic properties for GS could be attributed to the fact that these investigators used a mixture of glycosides, a fraction named GS_4, alcoholic extract of leaves and gymnemic acid. (Ricardo Galletto *et al., Braz. arch. biol. Technol*, 2004, 47(4), *Brazilian Archives of Biology and Technology*. On-line version ISSN 1678-4324.)

Heliotropium indicum Linn.

Hastishundi, Scorpion Tail

Toxicity of Pyrrolizidine Alkaloids

Several pyrrolizidine alkaloids have been isolated from this species. The active principle Indicine-oxide has reached Phase 1 clinical trials in advanced cancer patients. But severe toxic side-effects showed that a therapy with indicine-N-oxide was not justified. Most of the alkaloids are hepatotoxic and therefore internal use of *Heliotropium* species is not recommended. External application to promote wound healing and to fight infections seems to be less hazardous. (Dash KG, Abdullah SM, A Review on *Heliotropium Indicum, International Journal of Pharmaceutical Sciences and Research*, 2013, 4(4): 12531258.)

Inflorescence and seeds contain about 60 per cent pyrrolizidine alkaloids by the end of plant growth phase. N-oxides ranged between 63 and 90 per cent of total pyrrolizidine alkaloids, highest being in roots and inflorescence, leaves contain minimum amount. (*The Wealth of India, Second Supplement Series*, 2007, Vol. 2: 42.)

Antifertility and Abortifacient Activity

The petroleum ether extract of the entire plant is reported to possess significant antifertility activity when studied in rats. In another experiment, the n-hexane and benzene fractions of the ethanol extract of *Heliotropium indicum* were studied for antifertility activity in rats using anti-implantation and abortifacient models. *In vitro* sperm motility study was also performed using different concentrations of the extract. The study revealed that *Heliotropium indicum* possesses abortifacient activity and moderate effects on implantation and sperm motility. (Anupam Roy, *Journal of Pharmacognosy and Phytochemistry*, 2015, 4(3): 101-104.)

The aqueous and alcohol extracts of the plant possesses oxytocic activity. The roots contain significant amount of extradiol.

The extract of flowers in small quantities exhibited emmenagogue activity, also 40 per cent abortifacient activity in albino rats. (*The Wealth of India, First and Second Supplement Series*, 2002, 2007: 251 and 42 respectively.)

Holarrhena antidysenterica (L.) Wall.

Kutaja, Tellicherry Bark

Ineffective as a Single Drug

In a clinical trial, total alkaloids of the plant failed to cure amoebic affections of liver in a 16 year old male patient admitted at Carmichael Hospital for Tropical Diseases, Kolkata, suffering from irregular fever ushered in with rigor for two months. After 4 injections of the total alkaloids, the patient showed no improvement in fever and complained of severe pain in hypochondriac region, although he responded to emetine 1 grain *i.m.* (Chopra RN and Dey P, *Indian Med Gaz*, 1930, 65: 391.)

Toxic as a Single Drug

In a study undertaken at the Institute of Medical Sciences, BHU, Varanasi, the bark powder, 4 g/d in 3 divided doses for 15 days, produced side effects in almost 30 per cent (3 out of 11) patients of amoebiasis and giardiasis. The patients complainted burning sensation in head, abdomen and feet, dryness of mouth, nausea, agitation, nervousness, fatigue, insomnia, flatulence and constipation. 2 patients had vertigo while progressive hypotension was observed in 1 patient. Besides these, at 6 g/d dose, it produced acute hypotension and synocope in one male patient suffering from giardiasis and malabsorption syndrome. (Chaturvedi GN and Gupta JP, Side effects of traditional indigenous drug, Kutaja, *Indian J Physiol Pharmacol*, 27: 255-256.)

Effective when Used as an Additive to other Herbs

Kutajaghana Vati contains *Aconitum heterophyllium* Wall. Ex Royle (Atis) with *Holarrhena antidysenterica.* (The Ayurvedic Formulary of India. Part II: 175.)

Another Ayurvedic preparation, *Kutajaadi Vishesh Yoga*, in addition to *Aconitum heterophyllium,* contains Shyonaaka root bar, Bilva fruit pulp, Mocharasa, Mustaka root, Aamra seed kernel, Aardraka rhizome, Daadima flower buds and Lodhra bark. No single herb can get the credit of the curative effect of the compound drug.

Hyoscyamus niger Linn.

Paarsika-yavaani, Black Henbane

Poisoning when Used as an Intoxicant

Severe poisoning occur with the misuse of the drug as an intoxicant. Because of the high content of scopolamine in the drug, poisoning lead at first to somnolence, then after the intake of very high doses, to central excitation (restlessness, hallucination, deliria, and manic episodes), followed by exhaustion and sleep. Lethal doses carry with them the danger of asphyxiation (for adult starting at 100 mg atropine, with an alkaloid-rich drug at 30 mg, considerable less flor children.) Chief alkaloid (-)hyoscyamine, under storage conditions changes over to some extent into atropine and scopolamine. (*PDR for Herbal Medicines*, 2007: 439.)

Irrational Use in an Ayurvedic Medicine

Sarpgandhaaghan Vati, formulated by a contemporary *Vaidya* Yadavji Trikamji Acharya and approved by Ayurvedic Formulary of India, contains Sarpgandhaa root 537.7 mg, Paarsika-yavaani 107.14 mg, Jataamansi root 53.57 mg and *Cannabis sativa* leaves 57.57 mg in a pill (dose 750 mg to 1.25 g) for insomnia. If a patient takes a pill containing 1,125 mg, he will be consuming 178.56 mg of Paarsika-yavaani. What will be the synergistic action of Paarsika-yavaani with 3 other sedative drugs? It has been recorded that *Hyoscyamus niger* interacts with tricyclic antidepressants, amantadine, anti histamines, phenothiazines, procainamide and quinidine with concomitant use. (C. P. Khare.)

Illiciun verum Hook. f.

Takkola, Chinese Star Anise

Convulsive Fractions

Three new neurotropic sesquiterpenoids, veranisatins A, B and C, were isolated from star anise (*Illicium verum*.) Veranisatins showed convulsion and lethal toxicity in mice at a dose of 3 mg/kg (p.o.), and at lower doses they caused hypothermia. Veranisatin A and the related compound, anisatin, were tested for the other pharmacological activities such as locomotor activity and analgesic effect. Both compounds decreased the locomotion enhanced by methamphetamine at oral doses of 0.1 and 0.03mg/kg, respectively, and demonstrated the analgesia on acetic acid-induced writhing and tail pressure pain at almost similar doses. (Tomonori Nakamura *et al.*, Neurotropic Components from Star Anise (*Illicium verum*), *Chemical and Pharmaceutical Bulletin,* Nov 1996, 44(10): 1908-14. ResearchGate.)

In Faculty of Pharmaceutical Sciences, Chiba University, Japan, veranisatins were isolated in 1993 during a search on neurotropic components from medicinal plants. *Illicium verum* was observed to have hypothermic effect in mice and ethyl extract caused severe convulsions and lethally toxicity depending on dose. Oral administration of ethyl acetate extract after being defatted decreased the body temperature, caused severe convulsions and death in mice. One convulsive fraction was purified and two convulsants, veranisatin A (33 mg) and veranisatin B (20 mg) were isolated from 20 kg fruits of *I. verum*. Veranisatin A and veranisatin B presented convulsive effect and lethal toxicity (3/3) at a dose of 3 mg/kg p.o., while 1 mg/kg of both compounds showed hypothermic effect but no convulsions. (Emi Okuyama *et al.*, Convulsants from *Illicium verum*, *Chem Pharma Bull*, 1993, 41(9): 1670-1671.)

Illicium anisatum: The Adulterant of *I. verum*

Illicium anisatum (Shikimi, known as Japanese star anise) is similar to *I. verum* (Chinese star anise.) It is impossible to recognize Chinese and Japanese star anise in its dried or processed form. *Illicium anisatum* is not edible and highly toxic. Cases of illness, including serious neurological effects such as seizures, have been reported after using star anise tea, possibly due to the adulteration of Japanese star anise.

Japanese star anise contains anisatin, shikimin, and sikimitoxin, which cause severe inflammation of the kidneys, urinary tract, and digestive organs. Other compounds present in toxic species of *Illicium* are safrole and eugenol, which are not present in *I. verum* and are used to identify its adulteration. Shikimic acid is also present in the plant. Anisatin and its derivates are suspected of acting as strong GABA antagonists. The essential oil of air-dried *I. anisatum* obtained by hydrodistillation was analyzed by GC–MS. Fifty-two components were identified in the essential oil, and the main component was eucalyptol (21.8 per cent.) (http://eol.org)

Cases of product recalls have been reported when products containing star anise were found to be contaminated by Japanese anise. Cases of consumers admitted to hospital with neurological symptoms after ingesting excessive doses of star anise or smaller doses of products adulterated with Japanese anise were described in the literature.

Indigofera tinctoria Linn.
Nili, Indigo

Antidepressant and Noortropic Activity

A study was undertaken to evaluate the antidepressant and nootropic potential of aqueous extract of *Indigofera tinctoria* (ITAE.) The mice were randomly assigned into four groups (n=6) and they were pre-treated for 14 days before the experiment. Group I: Mice received the aqueous extract prepared for oral administration in distilled water using 0.2 per cent NaCMC as suspending agent, orally. Group II: Mice were treated with ITAE orally (250 mg/kg.) Group III: Mice were treated with ITAE orally (500 mg/kg.) Group IV: Mice were treated with imipramine orally (25 mg/kg) which served as Positive Control. For nootropic activity, mice were treated with piracetam orally (300 mg/kg) that served as positive control instead of imipramine.

Antidepressant activity was evaluated in mice by using the forced swim test (FST), tail suspension test (TST) and tetrabenazine induced catalepsy and ptosis models at doses of 250 and 500 mg/kg orally. Nootropic activity was assessed using elevated plus maze (EPM) Morris water maze. Imipramine (25 mg/kg) and piracetam (300 mg/kg) were used as standard drugs for antidepressant and nootropic activity respectively. Antioxidant assays like DPPH and TBARS was also carried out support the antidepressant and nootropic activity.

Pre-treatment with ITAE showed significant dose dependent reduction in immobility time displayed in both, FST and TST, when compared to that of vehicle. ($P<0.05$.) Also, significant reduction in tetrabenazine induced catalepsy and ptosis was observed. Moreover ITAE showed dose dependent cognitive enhancing activity in MWM and EPM. ITAE exhibited IC50 value of 17.39 µg/ml in DPPH assay and 398. 71 µgm/ml in TBARS assay. ITAE exhibits significant antidepressant and nootropic activities at the dose of 250 and 500 mg/kg which are comparable to imipramine and piracetam respectively. (Amrita A Choudhury and Archana R Juvekar, *International J Pharmacy Pharmaceutical Sciences*, 2014, Vol. 6(8): 131-135.)

Jatropha curcas Linn.

Dravanti, Physic Nut

Toxicity of Whole Seed as well as Detoxified Seed Meal in Livestocks

The seeds of *J. curcas* are known to be toxic in mice and rats. It has been reported that *J. curcas* seeds fed to the Nubian goats at doses ranging from 0.25 to 10 g/kg/day were found to be toxic with mortality occurred between 2 and 21 days. It was also observed that there was a decrease in glucose and marked rise in the concentration of arginase and glutamate oxaloacetate tranasaminase in the serum. Autopsy examination revealed hemorrhage in the rumen, reticulum, kidney, spleen and heart. Congestion, oedema of the lungs and excessive fluid in serous cavities were also noticed. Feeding mice as low as 1 mg/kg bw caused death.

The cake of *Jatropha curcas* after oil extraction is rich in protein and is a potential source of livestock feed. In view of the high toxic nature of whole as well as dehulled seed meal due to the presence of toxic phorbol esters and lectin, the meal was subjected to alkali and heat treatments to deactivate the phorbol ester as well as lectin content. After treatment, the phorbol ester content was reduced up to 89 per cent in whole and dehulled seed meal. Toxicity studies were conducted on male growing rats by feeding treated as well as untreated meal through dietary source. All rats irrespective of treatment had reduced appetite and diet intake was low accompanied by diarrhoea. The rats also exhibited reduced motor activity. The rats fed with treated meals exhibited delayed mortality compared to untreated meal fed rats (p 0.02.) There were significant changes both in terms of food intake and gain in body weight. Gross examination of vital organs indicated atrophy compared to control casein fed rats. However, histopathological examination of various vital organs did not reveal any treatment related microscopic changes suggesting that the mortality of rats occurred due to lack of food intake, diarrhea and emaciation. (K.D. Rakshit *et al.*, Toxicity studies of detoxified *Jatropha curcas* meal in rats, *Food and Chemical Toxicology*, 2006, 46: 3621-3625.*Cf.* ResearchGate.)

Juglans regia Linn.

Akshoda, Walnut Tree

The leaf and unripe fruit wax contain a toxic principle 5-hydroxy-1,4-napthaquinone (juglone) 28.6 and 29.8 per cent, respectively.

Effect on Skin and Mucous Membrane

Fresh walnut hull also contain juglone. Juglone acts as a mutagen in various model systems. Application of juglone-containing walnut preparations onto skin and mucous membranes leads to brown discoloration. The topical, daily use of juglone-containing preparations of walnut bark is tied to an increased occurrence of cancer of the tongue and leukoplakia of the lips. (*The Complete German Commission E Monographs, Blumenthal et al., American Botanical Council,* 1998: 381.)

Effect on Thyroid Hormones

The juice of unripe fruits showed significant thyroid hormone enhancing activity and may be used as a raw material for development of new therapeutic modalities against various thyroid diseases. However, prolonged use of such extract may cause serious effects. (*The Wealth of India, First Suppl. Series,* Vol. 4, 2003: 5.)

In Turkish folk medicine, the fruits and leaves of *Juglans regia* have been widely used as a herbal remedy for the treatment of various endocrine diseases such as diabetes mellitus, anorexia, thyroid dysfunctions. The effect of fruits of *J. regia* on the thyroid hormone levels of mice was investigated using two extracts prepared from the fruits by different methods. The acute toxicities of these two extracts in mice were assessed as well. On the basis of our findings obtained, the extracts prepared from the fruits of *J. regia* enhanced thyroid hormone levels, while they exerted minimal acute toxicity in mice. (S.Aydin *et al., Phytotherapy Research,* August 1994, 8(5): 308-310. ResearchGate.)

Effect on Human Fibroblasts

A study showed that juglone differentially reduces viability of human cells in culture. Normal fibroblast were found to be especially sensitive to juglone and lost viability primarily through a rapid apoptotic and necrotic response. This response may have been triggered by DNA damage since juglone induced a rapid and strong phosphorylation of H2AX in all phases of the cell cycle. Furthermore, juglone inhibits mRNA synthesis in human fibroblasts in a dose-dependent manner. Surprisingly, juglone caused a drastic reduction of the basal level of p53 in human fibroblasts and this loss could not be fully rescued by proteasome and calpain I inhibitors. However, when cells were pretreated with UV light or ionizing radiation, juglone was not able to reduce the cellular levels of activated p53. Our results show that juglone has multiple effects on cells such as the induction of DNA damage, inhibition of transcription, reduction of p53 protein levels and the induction of cell death. (Paulsen MT, Ljungman M, The natural toxin juglone causes degradation of p53 and induces rapid H2AX phosphorylation and cell death in human fibroblasts, *Toxicol Appl Pharmacol*, 2005 Nov 15, 209(1): 1-9.)

Juniperus communis auct. non Linn.

Hapushaa, Common Juniper

Hapushaa in *Ayurvedic Pharmacopoeia of India*, Part I, Vol. III, do not represent the unique profile of Juniper. Following text is based on adverse reactions of Juniper.

Antifertility Effect

The 50 per cent of ethanol extract of fruits (300 and 500 mg/kg p.o.) in rats revealed an antiplantation activity in a dose-dependent manner. The extract also showed abortifacient activity in both the doses when administered on day 14, 15 and 16 of pregnancy. The 90 per cent ethanolic extract of fruits induced dose-dependent (50-300 mg/kg) foetal resorption at early and late stages of pregnancy in rats. (*Cf. Reviews on Indian Medicinal Plants*, ICMR, Vol. 13: 492.)

Adverse Reactions

Single large doses of Juniper berries may cause catharsis, and repeated large doses may be associated with convulsions and renal damage. Kidney irritation from Juniper oil is examined in one study. It relates its effect to 1-terpinen-4-ol content. Products containing Juniper should be used with caution, and should never be used by those with reduced renal function.

Epidermal contact with Juniper Tar (preparation for psoriasis treatment) can cause potential carcinogenic DNA damage in human tissue. (*The Reviews of Natural Products*, Wolter Kluwer, 7th Edn: 880-881.)

A decoction of Juniper berries decreased glycemic levels in normoglycemic rats at a dose of 250 mg/kg by increasing peripheral glucose consumption and potentiating glucose-induced insulin secretion. Caution should be used with increased glucose levels in diabetics. (*PDR for Herbal Medicines*, 2007: 486-487.)

Lagenaria siceraria (Mol.) Standl.

Tumbi, Bitter Bottle-Gourd

Toxicity of the Juice

A 59 years old healthy male at Himalayan Institute Hospital, Dehradun, was reported to develop complaints of profuse bloody diarrhea, vomiting mixed with blood and oliguria within half an hour of consuming Bitter Bottle-Gourd juice. (Biochemical parameters returned to normalcy after 5 days treatment.)

Fifteen patients were reported to develop abdominal pain, vomiting, hematemesis, diarrhoea and hypotension within 15 min to 6 h after ingestion of Bitter Bottle-Gourd juice. Endoscopy showed esophagitis, gastric erosions, ulcer and duodenitis. (All patients recovered within one week.) (*Reviews on Indian Medicinal Plants*, ICMR, New Delhi, Vol.14, 2015: 39.)

An expert committee, constituted by the Department of Health Research at Indian Council of Medical Research (ICMR), on the basis of information collected from public domain, state and district health authorities, as well as from medical colleges, hospitals and private nursing homes across India, revealed that there were 3 deaths, one in Delhi and two in Uttar Pradesh, after consuming Bitter Bottle-Gourd juice or juice mixed with the juice of Bitter Gourd (*Momordica charantia.*) (They were over 59 years of age and had diabetes for 20 years.) Twenty six persons were admitted to various hospitals of the country on complaint of abdominal pain and vomiting following consumption of freshly prepared bottle gourd juice. Diarrhea and vomiting of blood was reported in 18 (69.2 per cent) and in 19 (73.1 per cent) patients, respectively. Biochemical investigations revealed elevated liver enzymes. More than 50 per cent patients had hypotension. Endoscopic findings showed profusely bleeding stomach with excessive ulceration in distal oesophagus, stomach and duodenum in most cases. (ICMR Task Force, Sharma S K, Puri R, Jain A, Sharma M P, Sharma A, Bohra S, Gupta Y K, Saraya A, Dwivedi S, Gupta K C, Prasad M, Pandey J, Dohroo NP, Tandon N, Sesikeran B, Dorle A K, Tandon N, Handa S S, Toteja G S, Rao S, Satyanarayana K, Katoch V M, *Assessment of effects on health due to consumption of Bitter Bottle-Gourd (Lagenaria siceraria) juice,* Indian J Med Res, 2012;135: 49-55.)

Protective role against thyrotoxicosis

At a dose of 100 mg/kg, peel extract inhibited the levels of serum thyroxine (T_4), triiodothyronine (T_3) and glucose as well as hepatic lipid peroxidation (LPO) concentration in mice. After 21 days of treatment a decrease in the concentration of thyroid hormones, glucose as well as LPO with parallel increase in antioxidants, indicated efficacy of test peel in amelioration of hypothyroidism, hyperglycemia and hepatic lipid peroxidation. (Dixit Y *et al., Int J Biomed Pharmaceut Sci,* 2: 78-83.)

The periplogenin-3-O-D-glucopyranosyl $(1 \rightarrow 6)(1 \rightarrow 4)$-D-cymaro-pyranoside, isolated from the fruits, was evaluated for L-thyroxine (L-T$_4$)-induced hyperthyroidism at three different concentrations (5, 10 and 25 mg/kg.) 5 mg/kg was found to be the most active dose as it could nearly normalize the level of T$_3$, glucose, insulin, Na$^+$-K$^+$ -ATPase activity, tissue LPO and different serum lipids suggestive of the protective role of periplogenin against thyrotoxicosis and associated cardiovascular problems. (Panda S. and Kar A., *Horm Metab Res,* 43: 188-193.)

Leucas cephalotes (Roth.) Spreng.

Dronpushpi (API), Gumma

Conflicting Findings for Plant's Hepatoprotective Activity

An earlier study on Gumma (common name of *Leucas cephalotes*) by Singh *et al.*, showed that the ethyl acetate extract of *Leucas cephalotes* (whole plant) failed to protect CCl_4 induced hepatotoxicity in mice and rats when used in a dose of 300 mg/kg, which was confirmed histologically. The extract also failed to reduce the CCl_4 induced mortality in 48 hr. The LD_{50} of this extract in mice was 1680 ± 21 mg/kg i.p. (Singh N *et al.*, An Experimental Evaluation of Protective Effects of Some Indigenous Drugs on Carbon tetrachloride induced Hepatotoxicity in Mice and Rats, *Quarterly Journal of Crude Drug Research*, 1978, 16: 8–16.)

Another study of Sharma *et al.*, showed that the 90 per cent ethanolic extract of the aerial parts of *Leucas cephalotes* was devoid of hepatoprotective activity. (Sharma ML *et al.*, Pharmacological Screening of Indian Medicinal Plants. *Indian Journal of Experimental Biology*, 1978, 16(2): 228-240.)

A recent study by G. Sofi *et al.*, was carried on hydro alcoholic (1: 1) extract (583 mg/kg, orally) which was higher than that used by Singh *et al.*, It was pointed out that ethyl acetate extract may have not possessed sufficient active chemical ingredients as water soluble ingredients might have been left out. It was concluded that Gumma (*Leucas cephalotes*) possesses significant degree of hepatoprotective effect against carbon tetrachloride-induced hepatotoxicity. (G. Sofi *et al.*, *Anc Sc Life*, Oct-Dec 2011, 31(2): 44-48.)

In any case, further studies are needed. *Plihaari Vatika* (*Bhaishajya Ratnavali*), for diseases of the spleen and liver, is a herbo-mineral compound processed in the plant juice of Dronpushpi.

Linum ustatissimum Linn.
Atasi, Flax

Flax Seed: An Alternative Source of New Drug for PCOS

A study at Chettinad Hospital and Research Institute, Kelambakkam, Tamilnadu, India, assessed the effects of flax seed powder on ovarian morphology, menstrual cycle, hirsutism and blood sugar in Polycystic Ovary Syndrome (PCOS.) 32 women with PCOS who fulfilled the selection criteria were included in this open label interventional study. Menstrual history, measurement of body weight, BMI, random blood sugar, ultrasound abdomen and Ferriman Galleway scoring for hirsutism were done at baseline. The subjects were given 15 grams of flax seed powder to be taken as a single dose in milk every morning for 3 months. 30 subjects completed the study. The above parameters were measured every month for three months. The results were analyzed using paired T test. A significant reduction in mean ovarian volume and number of follicles was observed after flax seed therapy. The mean reduction in right and left ovarian volume was -3.35 c.mm and -2.383 c.mm and the mean difference in number of follicles was -4.259, -4.519 for right and left ovaries respectively (p value <0.01.) After flax seed treatment 17 (56.7 per cent) subjects did not have peripheral follicles and 16 (53.3 per cent) had normal ovarian echogenicity. 12 subjects (40 per cent) had improved menstrual cycle and 3 (10 per cent) became pregnant. There was no significant change in hirsutism and blood sugar level after treatment. Flax seed supplementation has caused significant reduction in ovarian volume and number of follicles in polycystic ovaries, improved the menstrual cycles and not altered the body weight, blood sugar and hirsutism. This is probably is the first study to suggest that flax seed can be an alternative source of new drug for PCOS. (Fatima Farzana *et al., Int J Pharm Sci Rev Res*, March-April 2015, 31(1): 113-119.)

Advisory from National Institute of Health (NIH)

Women taking flax seed daily may experience changes in menstruation. The National Institute of Health (NIH), US department of health, advises that women with hormone-related health conditions be cautious about using flax seed. These conditions include endometriosis, polycystic ovary syndrome, uterine fibroids, and breast, ovarian or uterine cancer. Pregnant women also should not consume supplemental flax seed, as it could stimulate menstruation or cause other hormonal effects that might be harmful to the developing baby.

Omega-3 fatty acids can increase blood sugar levels, so people with diabetes should be cautious about consuming flax seed, according to the NIH. Flax seed also may elevate the risk of excessive bleeding because it decreases clotting. People with bleeding disorders or taking medications with blood-thinning effects may need to be cautious about using flax seed supplements. In bipolar disorders flax seed may cause mania.

Lobella inflata Linn.
Devanala, Indian Tobacco

The piperidine alkaloid lobeline was isolated as the main active component of *Lobelia*. Its absolute stereochemistry was determined in 1965. Related alkaloids include lobelanine and lobelanidine. These alkaloids are present in several other species of *Lobelia*.

Effect of Lobeline in Smoking Cessation

In nicotine pretreated rats, lobeline appeared to act as a short-acting antagonist of nicotine receptors, mediating the effects of nicotine on mesolimbic dopamine activity and locomotor stimulation. However, the effect of lobeline on N-methyl-D-aspartate serotonin and norepinephrine release have been reported to be complicated phenomena. Results from clinical studies of the efficacy of lobeline in withdrawal from tobacco smoking addiction have been mixed.

In clinical studies, doses of lobeline used for smoking withdrawal were 5 mg twice daily with 0.5 mg lozenges when there was an urge to smoke. Thus the therapeutic dose of lobeline is very close to the toxic dose. (*The Review of Natural Products*, Wolters Kluwer, 2012: 1002-1003.)

Toxicity of Overdose of Leaves

Overdose leads to nausea, vomiting, abdominal pain, diarrhea, burning in urinary passages, dizziness, headache, shivering, respiratory difficulties, paraesthesias, outbreak of sweating, bradycardia, cardiac arrhythmias, somnolence, and muscle twitching, respiratory failure. 0.6 to 1 g of the leaves are said to be toxic, 4 g fatal. (*PDR for Herbal Medicines*, 2007: 534.)

Malus pumila Mill.
Sinchitikaa, Apple

Cyanide Toxicity of Seeds

A small number of seeds found in one apple may be ingested without symptoms. Large amount of seeds exhibited potential toxicity. A man died from eating a cup of apple seeds, thinking that they were a delicacy. Because cyanogenic glycoside must be hydrolysed in the stomach in order to release cyanide, symptoms of poisoning occur after several hours. Apples contain amygdalin in their seeds, which is a cyanide-and-sugar based molecule. If the seed is chewed or otherwise broken, human or animal enzymes come into contact with the amygdalin and effectively cut off the sugar part of the molecule. The remainder can then decompose to produce the poisonous gas hydrogen cyanide.(*The Review of Natural Products*, Wolters Kluwer, 7th Edition, 2012: 93.)

In a study, the amygdalin content of apple seeds was found to be approximately 3 mg per g of seeds (one seed is approximately 0.7 g.) (Quantities of amygdalin in seeds of other fruits mg/g: apricot 14.4, red cherry 3.9, black cherry 2.7, peach 2.2, plum 2.2, pear 1.3.)

Episodes of Anaphylaxis after Ingestion of Apple

A 13-year-old boy with two episodes of anaphylaxis after exercise was suspected as food-dependent exercise-induced anaphylaxis (FEIAn.) The exercise challenge test after ingestion of an apple was performed, because apple was commonly contained in meals before the both episodes and a prick test for apple was positive. The exercise test was positive.

The prognosis of FEIAn has not been well described. Cases with FEIAn to apples like this case require a close follow-up in consideration of development of allergic reactions without exercise. (Mari Kaneko *et al.*, A case of apple allergy, *Japanese J Allergology*, 62(6): 698-703.)

Mentha viridis Linn.

Pudinah, Spearmint

Menthol Toxicity

Large doses of the peppermint oil can cause central nervous system stimulation; a dose of about 1 g/kg can be fatal. Menthol cannot be metabolized by neonates with Glucose-6-Phosphate Dehydrogenase (G6PD) deficiency; may cause jaundice. Inhalation of menthol or its instillation into the nose can cause nasal obstruction and apnea in infants. Allergy to mint oil can result in asthma. (M. Rotblatt, I. Ziment, *Evidence-based Herbal Medicine*, 2002: 275.)

Adverse Reactions

Loss of libido following consumption of peppermint tea has been reported in a study on the effects of peppermint teas on plasma testosterone, follicle-stimulating hormone, luteinizing hormone levels, and testicular issues in rats. (Akdogan M *et al.*, Urology, 2004, 6(2): 394-398.)

Rats fed peppermint oil in daily dosages up to 100 mg/kg for 28 days developed dose-related brain lesions. A similar 90-day study demonstrated no additional aggravation of cyst-like spaces in the cerebellum. (*The Review of Natural Products*, Wolters Kluwer, 7th Edition, 2012: 1258-1259.)

Interactions with Drugs

Absorption of a single dose of caffeine 200 mg was delayed by menthol 100 mg in healthy volunteers.

Peppermint oil 600 mg in water increased the area under the curve (AUC) of Felodipine. It elevated Felodipine plasma concentrations and increased pharmacologic effects and adverse reactions; also increased the AUC of Simvastatin 30 per cent in healthy volunteers. Decreased Cyclosporine levels were reported in patients consuming herbal tea containing peppermint and 8 other herbs. (*The Review of Natural Products*, Wolters Kluwer, 7th Edition, 2012: 1258-1259.)

Momordica charantia Linn.

Kaarvallaka, Bitter Gourd

Teratogenic and Spermatogenic Toxicity

Momorcharins, isolated from the seeds of the crude drug, have been shown to induce early and mid-term abortions in mice, and are teratogenic in cultured mouse embryos at the early organogenesis stage. Morphological abnormalities were observed in the head, trunk and limbs of the embryos. (*WHO monographs on selected medicinal plants*, Vol. 4; 205.)

Chronic administration (for 60 days) of the fruit extract (1.75 g/day) to dogs led to testicular lesions with mass atrophy of spermatogenic elements. (GV Satyavati, Ashok Gupta, Neeraj Tandon, Indian Council of Medical Research, *Medicinal Plants of India*, 1987, Vol. 2: 265.)

Effects on Hepatic Enzymes

The effect of the fruit on certain key hepatic enzymes was investigated in rats (daily dose: 10 ml of pure juice/kg bw, for thirty days.) Serum gamma-glutamyl transferase and alkaline phosphatase concentrations were found to be significantly elevated ($p < 0.001$ and $p < 0.01$-0.001, respectively.) Patients with liver disorders should seek advice from their health care professional before taking any preparation of the crude drug. Owing to reported adverse events such as hypoglycemia and convulsions in children, the crude drug and its preparations are contraindicated during breastfeeding. (*WHO monographs on selected medicinal plants*, Vol. 4: 201, 205.)

Fatal Outcome of Juice Therapy

An expert committee, constituted by the Department of Health Research at Indian Council of Medical Research (ICMR), on the basis of information collected from public domain, state and district health authorities as well as from medical colleges, hospitals and private nursing homes across India, revealed that there were 3 deaths, one in Delhi and two in Uttar Pradesh, after consuming bottle gourd juice or juice mixed with the juice of bitter gourd (*Momordica charantia.*) They were over 59 years of age and had diabetes for 20 years. (*Indian J Med Res*, 135: 49-55.)

Caution

Researchers have warned that *M. charantia* extract leads to a false negative test for sugar in urine (due to its ability to maintain the indicator dye in the glucose oxidase strips and the alkaline copper salts in a reduced state. (GV Satyavati, Ashok Gupta, Neeraj Tandon, Indian Council of Medical Research, *Medicinal Plants of India*, 1987, Vol. 2: 268.)

Moringa oleifera Lam.
Shigru, Drumstick Tree

Effect of Leaf on Thyroid Hormones

The role of *Moringa oleifera* aqueous leaf extract in the regulation of thyroid hormone, was studied in adult Swiss rats. Other than the thyroid hormone concentrations, hepatic lipid peroxidation (LPO) and the activities of antioxidant enzymes, superoxide dismutase (SOD) and catalase (CAT) were evaluated. In the first experiment, effects of the leaf extract (175 mg kg(-1) bw day(-1) for 10 days) were studied both in male and female animals. Following the administration of the extract, serum triiodothyronine (T_3) concentration and hepatic LPO decreased with a concomitant increase in the serum thyroxine (T_4) concentration, in female rats, while in males no significant changes were observed, suggesting that *Moringa oleifera* leaf extract is more effective in females than in the males.

To evaluate the impact of a higher dose, in the second experiment, the study was repeated in female rats, with 350 mg kg^{-1} bw day^{-1} for the same duration. Almost similar reduction in the serum T_3 concentration (approx. 30 per cent) and an increase in the T_4 concentration were observed suggesting the inhibiting nature of *Moringa oleifera* leaf extract in the peripheral conversion of T_4 to T_3, the principal source of the generation of latter hormone. As the antiperoxidative effects were exhibited only by the lower dose and percent decrease in T_3 concentration was nearly the same by both the doses, it is suggested that the lower concentration of this plant extract may be used for the regulation of hyperthyriodism. (Tahiliani P, Kar A, *Pharmacol Res.* 2000 Mar; 41(3): 319-23. PMID 10675284.)

Mucuna pruriens Hook.

Kapikacchu, Cowhage

Cases of Acute Toxic Psychosis

In the remote Memba district of Mozambique severe famine occurred because of drought and civil war in 1989, and over a 6-week period 203 cases of acute toxic psychosis, mainly in women and girls, were reported. They were confused and agitated, with headache, palpitations, hallucinations and paranoid delusions. They recovered rapidly after treatment with intravenous chlorpromazine. The disease was attributed to eating the seeds of a wild variety of *Mucuna pruriens*. The seeds contain levodopa, *N,N*-dimethyltryptamine and its 5-methoxy derivative, bufotenine and other alkaloids. In times of famine it is eaten after detoxification by repeated boiling in water, which is thrown away. Shortage of water may have led to insufficient treatment; women may have been the principal victims. (Infante ME *et al.*, Outbreak of acute toxic psychosis attributed to *Mucuna pruriens*, *Lancet*, 1990 Nov 3; 336(8723): 1129. PMID: 1978001.)

Abuse as a Drug

Properties of *Mucuna pruriens* are a result of its contents of natural L-dopa, a direct precursor to the neurotransmitter dopamine. As the nervous system needs higher and higher levels of dopamine to produce the same response, an individual seeks out more and more of the stimulating substance. This leads to addiction.

Altering the level of brain chemicals like dopamine and serotonin also effect many other chemicals, hormones and enzymes that keep the body in balance. The long term impacts on healthy humans taking high doses of *Mucuna pruriens* in sex tonics or for its L-dopa content warrant further research. It is also not known whether L-dopa is toxic to dopamine neurons. (Leslie Taylor, *The Healing Power of Rainforest Herbs*, 2005, SquareOne: 444.)

Myristica fragrans Houtt.

Jaatiphala, Nutmeg

Effect of Nutmeg on the Kidneys

Adult Wister rats of both sexes (n=24), with average weight of 220 g were randomly assigned into two treatments (A and B) of (n=16) and Control (C) (n=8) groups. The rats in the treatment groups (A and B) received 0.1g (500 mg/kg bw) and 0.2 g (1000mg/kg bw) of nutmeg thoroughly mixed with the feeds respectively on a daily basis for forty-two days. The control group (C) received equal amount of feeds daily without nutmeg added for forty-two days. The rats were given water liberally. The rats were sacrificed by cervical dislocation on the forty-third day of the experiment. The kidneys were carefully dissected out and quickly fixed in 10 per cent buffered formaldehyde for routine histological study after hematoxylin and eosin method. The histological findings in the treated sections of the kidneys showed distortion of the renal cortical structures, vacuolations appearing in the stroma and some degree of cellular necrosis, with degenerative and atrophic changes when compared to the control group. These findings indicate that oral administration of nutmeg may have some deleterious effects on the kidneys of adult Wistar rats at higher doses and by extension may affect its excretory and other metabolic functions. It is recommended that dosage recommended by the healthcare provider should must be followed. (Andrew Osayame Eweka and Abieyuwa Eweka, *N Am J Med Sci*, 2010 Apr, 2(4): 189-192.)

Poisoning Due to Cuncurrent Use of Nutmeg and Flunitrazepam

In literature, cases of nutmeg abuse have been described repeatedly, but only one fatal case of poisoning was reported. In the reported case, myristicin (4 mcg/ml) was detected for the first time in the postmortal serum of a 55-year-old woman. Identification was achieved with the aid of UV-VIS spectroscopy and TLC; for quantification, HPLC was used. Because also flunitrazepam (0.072 microg/ml) was found, death had probably been due to the combined toxic effect of both substances.

From 1996 to 1998, in a series of cases, seven poisonings with nutmeg were recorded by the Erfurt Poison Information Centre. Even where higher doses (20-80 g of powder) had been ingested, a life-threatening situation was never observed. In one of these cases, myristicin blood level of 2 mcg/ml was measured 8h after ingestion of two to three tablespoonful of nutmeg powder (approx. 14-21 g, or 280-420 mg/kg.) (Stein U *et al.*, *Forensic Sci Int*, 2001 Apr 15, 118(1): 87-90.)

Nardostachys jatamansi DC.
Jatamansi, Spikenard

Antiandrogenic Activity

In a study, plants were screened for anti-androgenic activity using the RIKILT yeast Androgen bioAssay (RAA.) Selected positive plants were subsequently tested for their efficacy against PCOS induced by estradiol valerate (EV) in rat models. RAA revealed the antiandrogenic property of *Nardostachys jatamansi*, and *Tribulus terrestris*. EV administration reduced the weight gain and disrupted cyclicity in all rats. *Nardostachys jatamansi* and *Tribulus terrestris* extract treatment normalized estrous cyclicity and steroidal hormonal levels and regularized ovarian follicular growth. The *in vitro* antiandrogenic activity of plant extracts and their positive effects on different parameters of PCOS were proved *in vivo*. (Palakkil Mavilavalappil Sandeep, Toine FH Bovee, and Krishnan Sreejith, Metabolic Syndrome and Related Disorders, published in Volume 13, Issue 6, July 20, 2015. Online ahead of Print: April 28, 2015.)

Effects on Rat Brain

The effect of acute and subchronic administration of an alcoholic extract of the roots of *Nardostachys jatamansi* on norepinephrine (NE), dopamine (DA), serotonin (5-HT), 5-hydroxyindoleacetic acid (5-HIAA), gamma-aminobutyric acid (GABA), and taurine were studied in male albino Wistar rats. The acute oral administration of the extract did not change the level of NE and DA but resulted in a significant increase in the level of 5-HT and 5-HIAA. A significant increase in the level of GABA and taurine was observed in the drug-treated groups when compared to the controls. A 15-day treatment resulted in a significant increase in the levels of NE, DA, 5-HT, 5-HIAA, and GABA. These data indicate that the alcoholic extract of the roots of *N. jatamansi* causes an overall increase in the levels of central monoamines and inhibitory amino acids. (Prabhu V *et al., Planta Med*, 1994 Apr, 60(2): 114-117.)

Nerium oleander Linn.

Karavira, Red Oleander

Nerium indicum differs from *Nerium oleander* only in bearing fragrant flowers. *N. indicum* bears white and red flowers. Yellow-flowered species is equated with *Thevetia neriifolia*. Toxic manifestations of Yellow Oleander also include cardiac dysrhythmias. In *Ayurvedic Pharmacopoeia of India*, Vol. 1, *N. indicum* and *N. oleander* are synonyms.

Toxicity of Leaf

Non-lethal dose of 70 per cent ethanol extract of the *Nerium oleander* dry leaves (1000 mg/kg body weight) was subcutaneously injected into male and female mice once a week for 9 weeks (total 10 doses.) One day after the last injection, final bw gain (relative percentage to the initial body weight) had a tendency, in both males and females, towards depression suggesting a metabolic insult at other sites than those involved in myocardial function. Multiple exposure of the mice to the specified dose failed to express a significant influence on blood parameters (WBC, RBC, Hb, HCT, PLT) as well as myocardium. On the other hand, a lethal dose (4000 mg/kg bw) was capable of inducing progressive changes in myocardial electrical activity ending up in cardiac arrest. The electrocardiogram abnormalities could be brought about by the expected Na^+, K^+-ATPase inhibition by the cardiac glycosides (cardenolides) content of the lethal dose. (Haeba MH *et al.*, Toxicity of *Nerium oleander* leaf extract in mice, *J Environ Biol*, 2002, Jul 23(3): 231-237.)

Case Reports

A 21-year-old female was admitted in the emergency room with vomiting and lightheadedness 15 hours after ingestion of common *Oleander,* aqueous leaf extract (10-20 leaves.) She had been advised to take the extract in order to conceive a baby.

On initial examination, the blood pressure was 122/80 mmHg with irregular pulse of 46/min. She was looking toxic due to excessive vomiting. Other general physical parameters were normal. Cardiovascular examination revealed an irregular rhythm with soft S1 and normal audible S2 over the cardiac apex. Electrocardiogram revealed inverted P wave in inferior lead and prolonged PR interval (.28 s), with varying degree AV blocks and normal QRS duration.

Thus the most serious side effects of *Oleander* poisoning are cardiac abnormalities, including various ventricular dysrhythmias, tachyarrhythmias, bradycardia, and heart block. (Ibraheem Khan *et al., Heart Views.* 2010 Oct-Dec,11(3): 115–116.)

In a patient, who ingested *Oleander,* the serum digoxin levels were found high (4.4 ng/mL) and were associated with bradyarrhythmias and tachyarrhythmias, which decreased as the serum concentration of the toxin decreased. Another patient who ingested 7 *Oleander* leaves had digoxin serum levels of 5.69 nmol/L.

In birds, as little as 0.12 to 0.7 g of the plant has caused death. As few as 15 to 20 g of fresh leaves can be fatal to a horse. The flower nectar makes honey toxic. (*The Review of Natural Products*, Wolters Kluwer, 7th Edition, 2012: 1191.)

For Further Research

Recently, research has focused on anticancer effects of *Oleander*. Oleandrin inhibited certain kinases, transcription factors, and inflammatory mediators, including tumor necrosis factor. (Manna SK *et al., Cancer Res*, 2000, 82: 97-103.)

Contraindications

Contraindicated with Digoxin/cardiac glycosides.

Nigella sativa Linn.
Upkunchikaa, Kalanji

The Effects of Thymoquinone

Experimentally, it has been demonstrated that *N. sativa* extracts and the main constituent of their volatile oil, thymoquinone, possess antioxidant, anti-inflammatory and hepato-protective properties. To further evaluate the toxicological properties in a metabolically competent cellular system, thymoquinone was applied to primary rat hepatocyte cultures, and both cyto- and genotoxic effects were tested. Mitotic indices and the rates of apoptoses and necroses were determined as endpoints of cytotoxicity, while chromosomal aberrations and micronucleated cells served as endpoints of genotoxicity. In this approach thymoquinone demonstrated cyto- and genotoxic effects in a concentration dependent manner: it induced significant anti-proliferative effects at 20 microM and acute cytotoxicity at higher concentrations. Thymoquinone significantly increased the rates of necrotic cells at concentrations between 2.5 and 20 microM. Furthermore, it induced significant genotoxicity at concentrations > or =1.25 microM.

These observations support the previous finding that thymoquinone causes glutathione depletion and liver damage, but contradict the reports indicating antioxidant and anti-clastogenic effects. Thymoquinone might be metabolised to reactive species and increase oxidative stress, which contributes to the depletion of antioxidant enzymes and damage to DNA in hepatocytes treated with high thymoquinone concentrations. (Khader M *et al., In vitro* toxicological properties of thymoquinone, *Food Chem Toxicol,* 2009 Jan, 47(1): 129-33.) (PMID: 19010375.)

Ocimum sanctum Linn.

Tulasi, Holy Basil

Adverse Effects on Hormonal Functions

Intragastric administration of the leaves prevented implantation of embryo in various animal models. Intragastric administration of the leaves (10 per cent of feed) to male mice inhibited spermatogenesis. (*WHO monographs on selected medicinal plants*, Vol. 2; 213.)

Treatment with benzene extract of the leaves at 150, 200 and 250 mg/kg *p.o.* for 15 days showed dose-dependent antiandrogenic activity by reducing sperm count, number of motile sperms, forward velocity and the increased number of abnormal sperms. It also showed disturbance in the proteins and alteration in cauda epididymal milieu. (Ahmad M. *et al.*, *Oriental Pharm Exp Med*, 2009, 9(4): 339-349.)

Treatment of albino rats with a benzene extract of *Ocimum sanctum* leaves (250 mg/kg body weight) for 48 days decreased total sperm count, sperm motility, and forward velocity. The results suggest that such effects are due to androgen deprivation, caused by the anti-androgenic property of *Ocimum sanctum* leaves. The effect was reversible because all parameters returned to normal 2 week after the withdrawal of treatment. A significant decrease was noted in the sperm count in rabbits. Serum testosterone levels showed marked increase while FSH and LH levels were significantly reduced in *Ocimum sanctum*-treated rabbits (2 g fresh leaves/rabbit for 30 days.) (*International journal of Ayurveda research*, October 2010, 1(4): 208-10.)

Benzene and petroleum ether extracts of the plant have also shown 80 per cent and 60 per cent antifertility respectively in female rats. On feeding the male albino mice with leaves, biochemical changes were observed, indicating the sterility of the treating mice. But the use of *Ocimum sanctum* for fertility regulation may prove futile since it depresses the mating response of the treated animals and does not bring about azoospermia. (*The Wealth of India, Second Supplement series*, Vol.2: 232.)

Effect on Thyroid and Endogenous Antioxidant Enzymes

The effects of *Ocimum sanctum* leaf extract on the changes in the concentrations of serum triiodothyronine (T_3), thyroxine (T_4) and serum cholesterol; in the activities of hepatic glucose-6-phosphatase (G-6-P), superoxide dismutase (SOD) and catalase (CAT); hepatic lipid peroxidation (LPO) and on the changes in the weight of the sex organs were investigated. While the plant extract at the dose of 0.5 g kg^{-1} body wt. for 15 days significantly decreased serum T_4 concentrations, hepatic LPO and G-6-P activity, the activities of endogenous antioxidant enzymes, SOD and CAT were increased by the drug. However, no marked changes were observed in serum T_3 level, T_3/T_4 ratio and in the concentration of serum cholesterol. It appears that *Ocimum sanctum* leaf extract is antithyroidic as well as antioxidative in nature. (Panda S and Kar A, *Ocimum sanctum* leaf extract in the regulation of thyroid function in the male mouse, *Pharmacol Res.*, 1998 Aug, 38(2): 107-110.)

Papaver somniferum Linn.

Ahiphena, Opium Poppy

Effect on Sex Hormones

A reduction in the level of sex hormones has been observed in survivors of cancer who have consumed opioids for long periods. This has been associated with significantly higher levels of depression, fatigue, and sexual dysfunction.

Opioid induced androgen deficiency (OPIAD) has recently been identified in men taking long-term opioids through the oral, intrathecal and transdermal routes. OPIAD is a clinical condition characterized by decrease of testosterone by 39 per cent. These women were not taking estrogen supplements, indicating that there is an impairment of adrenal androgen production.

Animal studies suggest that opioid induced hypogonadism may be due to suppression of hypothalamic, pituitary and testicular function.

A recent review highlighted the impact of opioids on the endocrine system. The authors concluded that patients on longterm opioid therapy should be screened routinely for symptoms of hypogonadism and laboratory abnormalities in sex hormones. (Raghavan S *et al., Trends in anaesthia and critical care*, 1: 2011. *Science Direct*, www. Elsvier.com) Also see: Daniell HW, Hypogonadism in men consuming sustained-action oral opioids, *J Pain*, 2002, Oct, 3(5): 377-384.)

Pergularia daemia (Forsk.) Chiov.

Vishaanikaa, Hairknot Plant

Antifertility Activity of *Pergularia daemia*

An earlier report has shown that the C_2H_5OH extract of *Pergularia daemia* and its steroidal fraction possess significant anti-implantation and abortifacient activity. This study is an attempt to evaluate the antifertility activity of the alkaloidal fraction and to compare its activity with that of the steroidal fraction.

Adult albino mice (20 female and 10 male) of proven fertility were used in the study. The female and male mice were caged separately and maintained on standard balance diet. Before initiation of antifertility activity experiments, the female mice having regular oestrous cycle were selected and kept in cages in a ratio of two female: one male for mating. The vaginal smear of the female mice was examined daily for the presence of thick cluster of spermatozoa. The day, on which thick cluster of spermatozoa was found in the vaginal smear, was marked as day one of pregnancy and males were withdrawn from the cages on the day. The female mice received drug through oral route for 1 to 9 days of pregnancy. The animals of the control group received the vehicle only. Animals were laparotomized on the 10th day and numbers of implantation sites were observed in the horns of the uterus.

Since both the fractions of the C_2H_5OH extract at a dose of 200 mg kg^{-1} bw inhibited the implantation of the fertilized ova, the activity of the alkaloidal fraction was compared with that of the steroidal fraction in terms of their effect on the oestrous cycle. The oestrous cycle of the alkaloidal fraction treated mice returned to normal from pregnancy within 4 to 6 days of drug treatment. On the other hand the oestrous cycle of the steroidal fraction treated mice returned to normal within 6 to 8 days of drug treatment. This suggests that the alkaloidal fraction is more active and possesses no effect of prolonging the oestrous cycle compared to that of steroidal fraction. The alkaloidal fraction of the ethanol extract has great potential to prevent fertilization in the female mice. (Md. Golam Sadik *et al., Journal of Medical Sciences*, 2001, 1: 217-219.)

Piper betle Linn.

Naagavalli, Betle Pepper

Antiestrogenic, Antifertility Effect

A recent (2014) study showed the antiestrogenic, antifertility effect of leaves of *Piper betel* (PBL) in female albino rats. The estrous cycle was irregular and prolonged in the treated groups indicating anestrus condition, which would result in infertility. Both aqueous and methanolic extracts showed a significant decrease in the duration of proestrus and estrus with a prolonged diestrus at 1000 mg/kg/day and 1500 mg/kg/day doses as compared with control. However, no change was seen in the metestrus phase. The rats treated with PBL showed a significant ($P < 0.05$), dose-dependent decrease in the estrus phase, in comparison to the control group, the effect was more with the methanolic extract. Large, cornified cells appeared after proestrus phase with decreased number of cornified cells. There was a significant reduction in the number of the estrous cycle, in the PBL treated group. Anestrus phase appeared in all the rats treated with the aqueous and methanolic PBL extract, which was not observed in the control group. However, the aqueous extract at a dose of 500 mg/kg/day had no effect either on the estrous cycle or on its different phases. The observed effect of PBL could be due to the flavonoids and saponin contents, which also contributes to its antiestrogenic mechanism of action. (Sasmita Biswal, Phytochemical analysis and a study on the antiestrogenic antifertility effect of leaves of *Piper betel* in female albino rat, *Anc Sci Life*, 2014 Jul-Sep; 34(1): 16–22.)

Plumbago indica Linn.
Rakta Chitraka, Red Lead wort

Plumbagin-Induced Hepatotoxicity

In a study, the effects of plumbagin and *Plumbago indica* extract (PI) on hepatic histomorphology and antioxidative system in mice were examined.

Adult male intelligent character recognition mice were intragastrically administered plumbagin (1, 5, and 15 mg/kg/day) or PI (20, 200, and 1,000 mg/kg/day) consecutively for 14 days. Hepatic histomorphology was examined. Plasma alanine transaminase (ALT) and aspartate transaminase (AST) levels, hepatic lipid peroxidation, superoxide dismutase (SOD), catalase (CAT), and glutathione peroxidase (GPx) activities, and the ratio of reduced to oxidized glutathione (GSH/GSSG) were determined.

Plumbagin and PI concentration-dependently induced hepatic injury based on histopathological changes via imbalance of antioxidative system. Plumbagin and PI significantly increased plasma ALT and AST levels, hepatic lipid peroxidation, and GPx activity but significantly decreased hepatic SOD and CAT activities. The GSH/GSSG ratio was significantly reduced by plumbagin.

Plumbagin and PI caused hepatotoxic effects in the mice by unbalancing of the redox defense system. Therefore, plumbagin and PI-containing supplements should be used cautiously, especially when consumed in high quantities or for long periods. (Nadta Sukkasem *et al., J Intercult Ethnopharmacol*, 2016 Mar-Apr, 5(2): 137-145.)

Effect of Plumbagin-Free Alcohol Extract

In an experimental study, the effect of plumbagin-free alcohol extract (PFAE) of *Plumbago zeylanica* root, on female reproductive system and fertility of adult female wistar rats was assessed.

After the oral acute toxicity study, the PFAE was administered at two dose levels to perform the estrous cycle study, anti-implantation and abortifacient activity and hormonal analysis. However, the estrogenic/antiestrogenic activity was evaluated at only one most effective dose. LD_{50} cut-off was 5,000 mg/kg body weight. The extract exhibited significant anti-implantation and abortifacient activity at the tested dose levels (300 and 500 mg/kg, p.o.) ($P<0.01$.) The extract dose-dependently decreased the levels of serum progesterone, follicle stimulating hormone and luteinizing hormone, while a dose-dependent increase was observed in the concentration of serum prolactin. The extract did not show any significant changes in structure and function of uterus when given alone, but when given along with ethinyl estradiol, it exhibited significant antiestrogenic activity in immature overiectomized female rats ($P<0.001$.)

Biochemical parameters in the serum/blood and haematological parameters did not show appreciable changes throughout and after the course of investigation. However, all the altered parameters returned to normalcy within 30 days following withdrawal of treatment. (Sandeep G *et al., Asian Pac J Trop Med*, 2011 Dec;4(12): 978-84. Hainan Medical College, Haikou, China.)

Effect of Plumbagin on Male Dog Reproductive System

The role of plumbagin on male reproductive function was studied in relation to its antifertility activity. It caused testicular lesions and increased serum cholesterol, suggestive of phytoesterogen like activity. (Bhargava SK, Effects of plumbagin on reproductive function of male dog, *Indian J Exp Biol*, 1984 Mar, 22(3): 153-156.)

Psidium guajava Linn.
Peruka, Guava

Inotropic Effect of an Ethanol Extract of the Leaves

In guinea-pig atria, an ethanol extract of the leaves reduced atrial contractions by depressing the myocardial force in a concentration-dependent manner (median effective concentration (EC_{50}=1.4 g/l.) Concentrations higher than 2.5 g/l completely abolished the myocardial contractility. Further more, an acetic acid fraction (EC_{50} =0.07 g/l) of the extract increased the relaxation time measured at 20 and 50 per cent of the force curve by 30 and 15 per cent, respectively, but did not change the contraction time. The negative inotropic effect of the extract was abolished by atropine sulfate, suggesting that either the active substance acts as a cholinergic agonist or that it could release acetylcholine from parasympathetic synapses. (*WHO monographs on selected medicinal plants*, Vol. 4: 133.)

Effect against Caffeine-Induced Spermatotoxicity

A recent study investigated the protecting potential of Aqueous Guava (*Psidium guajava*) Leaf Extract (AGLE) against caffeine induced spermatotoxicity in albino rat models. Thirty healthy and sexually matured albino rats were divided into five groups of six rats each using a completely randomized design. They were treated with caffeine and AGLE combinations orally for 65 days. The result showed that caffeine significantly ($p<0.05$) reduced sperm viability, sperm count and sperm motility, while sperm head abnormality increased in caffeine treated rats when compared to the control. However, AGLE significantly ($p<0.05$) protected the treated albino rat models from caffeine induced spermatotoxicity in a dose-dependent manner. These results show that AGLE is effective in protecting albino rat models against caffeine induced spermatotoxicity in a dose dependent manner. (Utip Ekaluo *et al.*, *Research Journal of Medicinal Plant*, January 2016, 10(1): 98-105.)

Pueraria tuberosa DC.
Vidaarikanda, Indian Kudzu

The Role of Phytoestrogenic Compounds in Male Fertility

A study investigated the effects of ethanolic extract of *P. tuberosa* (PT) on sexual behaviour and androgenic activity. Male albino rats were divided into four groups of six animals each: control group 1 (2 per cent acacia solution), PT-treated group 2 (50 mg/kg), PT-treated group 3 (100 mg/kg), and PT-treated group 4 (150 mg/kg.) The extract treatment for 28 days to rats significantly improved the androgenic and sexual behaviour parameters. There was also an increase in serum concentration of FSH and improvement in serum testosterone level in group treated with PT. Administration of PT showed a significant androgenic stimulation as evidenced by an increase in the weights of the testis, epididymis, and seminal vesicles. Spermatogenesis was also improved and is evidenced by improvement in the histoarchitecture of testicular sections.

In one of the allied species of PT, *Pueraria mirifica*, containing phytoestrogens similar to PT, there was an increase in the level of LH and FSH along with an improved mating efficiency. A number of scientific investigations have shown that phytoestrogen compounds exert biological activity via the central nervous system. Amongst other pharmacological properties of phytoestrogens are their antioxidant, neuroprotective, antidepressant, and anxiolytic activities. Phytoestrogens present in PT might be contributing to the improvement of sexual behaviour in rats. Phytoestrogens like daidzein and genistein also affect neurobehavioural actions are largely antiestrogenic, either antagonising or producing an action in opposition to that of oestradiol. The extract might be acting through is the stimulation of endogenous estrogens synthesis which is a contributing factor in male fertility. Increases in LH, FSH, and testosterone levels also indicate an effect of PT on gonadotropin-releasing hormone (GnRH.) GnRH agonist effect may be the mechanism involved in the androgenic and estrogenic activities evidenced in male rats. (Chauhan NS *et al., Scientific World Journal*, Dec 30, 2013: 780659. Published online 2013 Dec 30. doi: 10.1155/2013/780659.)

Punica granatum Linn.

Dadima, Pomegranate

Toxicity of Dried Root or Trunk Bark

Dried root or trunk bark is used orally to treat dyspepsia, sore throat, menorrhagia, leukorrhoea and ulcers (3-9 g for decoction.)

Ingestion of 80.0 g of dried root or trunk bark may cause severe vomiting with blood, dizziness, fever, tremor, and collapse. After 10 hours to 3 days, temporary blindness may occur, which usually resolves after several weeks.

Ingestion of pelletierine may cause visual disturbances with dilated pupils, dizziness and headache, as well as long-lasting anaesthesia or somnolence. Further symptoms of overdose include colic, cold sweat, dizziness, headache, muscle cramps, weakness or paralysis of lower extremities, nausea, cardiac and respiratory collapse. (*WHO monographs on selected medicinal plants*, Vol. 4: 113.)

(The rind of the stem contain 0.4 per cent piperidine alkaloids, and up to 0.8 per cent in the rind of the root. (*PDR for Herbal Medicines*, 2007: 663.)

Interaction with Rosuvastatin

In one case, a patient taking rosuvastatin 5 mg every other day in combination with ezetimibe 10 mg daily, developed rhabdomyolysis 3 weeks after starting to drink pomegranate juice 200 ml twice weekly. This patient had a history of elevated creatine kinase levels while not receiving any statin treatment. (*Natural Medicines Comprehensive Database*, 2013: 1265.)

Randia spinosa Poir.

Madanaphala, Common Emetic Nut

Toxicity of the Fruit Pulp

The ethanolic extract of the pulp showed stimulant action on isolated guinea-pig uterous. In experimental animals, crude saponin produced salivation; on contact caused a generalized irritation of the mucous membranes producing sneezing and vomiting, and bleeding from urinary tract. The cornea was inflamed and the drug caused haemolysis, both *in vitro* and *in vivo*. The perfused frog heart was arrested in a few minutes; with higher concentrations practically instantaneously. The drug was rapidly detoxified in the liver. (*The Wealth of India*, Vol. VIII, original edn., 1969: 362.)

The defatted pulp produced CNS depressant effect and potentiates the hypnotic effect of pentobarbitone in rats. It depresses heart, relaxes ileum, and antagonises effect of acetyl choline. These effects were transitory. (*The Wealth of India, First Supplement Series*, 2004, Vol 5: 1.)

Rauvolfia serpentina (L.) Benth. ex Kurz
Sarpagandhaa, Indian Snakeroot

Rauvolfia serpentina is not a classical Ayurvedic drug. During classical period, Dalhan's Sarpachhatrikaa and Varshaasu chhatrakaaraa, and other synonyms, Naakuli, Gandha Nakuli, Sarpa-gandhini, Sarpa-netra indicate that Sarpagandhaa of Ayurvedic period was a toxic mushroom (Kavaka.)

Rauvolfia serpentina was identified by German scientist Rauwolf. It was first recorded in Europe in 1785, but its efficacy was not screened until 1946. It was tried in modern medicine to lower blood pressure and for controlling symptoms of mental illness. The drug is now used only sparingly.

Adverse Side Effects of Reserpine

R. serpentina was used by many physicians throughout India in the 1940s and then was used throughout the world in the 1950s, including in the United States and Canada. It fell out of popularity when adverse side effects, including depression, became associated with it.

From 1959 to 1960, 151 cases of toxicity were reported in the United States from consuming *Rauvolfia*, and only 4 per cent of these cases were in adults. Nausea, vomiting, hypotension, sedation, and coma have been described by patients. Also symptoms of bradycardia and facial flushing were reported. Psychiatric depression was most common with doses of reserpine of greater than 0.5 mg per day and was significantly decreased in a daily dose of less than 0.25 mg of reserpine. Between 1962 and 1965, 225 reports of accidental ingestion were reported in the United States. Three cases were reported of children between the ages of 30 months and 4 years who ingested reserpine in doses as high as 25 mg. All cases were resolved. (Douglas Lobay, *Rauvolfia* in the Treatment of Hypertension, *Integr Med (Encinitas)*, 2015 Jun, 14(3): 40–46.)

Rauvolfia serpentina is a safe and effective treatment for hypertension. The plant was used by many physicians throughout India in the 1940s and then was used throughout the world in the 1950s, including in the United States and Canada. It fell out of popularity when adverse side effects, including depression and cancer, became associated with it. This author reviews the scientific literature with regard to the use of *Rauvolfia* and the treatment of hypertension. The author reviews the plant's botany, chemistry, and pharmacology and provides a researched and documented method of action for the active ingredients. With special emphasis on the plant's role in treating high blood pressure, the author looks at medical uses of the plant, critically examining its adverse side effects, toxicology, and carcinogenicity. The author refutes the association between the plant and carcinogenicity and discusses the importance of correct dosing and of screening patients to minimize the occurrence of depression. He concludes with the recommendation of use of low dose *Rauvolfia* (LDR) for suitable patients with hypertension. The plant provides clinicians with a safe and effective adjunct to pharmaceuticals in the treatment of high blood pressure. Adverse side effects of reserpine included lethargy, sedation, psychiatric depression, hypotension, nausea, vomiting, abdominal cramping, gastric ulceration, nightmares, bradycardia, angina-like symptoms, bronchospasm, skin rash, itching, galactorrhea, breast enlargement, sexual dysfunction, and withdrawal psychosis in 1 case. The most common side effect noted in all patients was nasal congestion, occurring in 5 per cent to 15 per cent of all patients. After several months of use, mental depression can occur and may persist. With extremely large doses, Parkinson-like symptoms, extrapyramidal reactions, and convulsions can occur. Allergic reactions to *Rauvolfia*, including asthma, are rare.

Adequate doses of reserpine that produce decreased blood pressure will not cause reserpine-induced gastric ulcerations. Reserpine has been observed to cause a slight edema in some patients. Possible interactions with other drugs include cardiac glycosides, ephedra, alcohol, antipsychotic drugs, barbiturates, digoxin, diuretics, ephedrine, levodopa, monamine oxidase inhibitors, propranolol, stimulant drugs, and tricyclic antidepressants. *Rauvolfia* may interact with the following lab tests, including tests for corticosteroids, bilirubin, catecholamines, gastric acidity, norepinephrine, prolactin, thyroxine, and vanillylmandelic acid. (Douglas Lobay, *Rauvolfia* in the Treatment of Hypertension, *Integr Med (Encinitas)*, 2015 Jun, 14(3): 40–46.)

Irrational Use in an Ayurvedic Medicine

Sarpgandhaa-ghan Vati, formulated by a contemporary *Vaidya* Yadavji Trikamji Acharya and approved by Ayurvedic Formulary of India, contains Sarpgandhaa root 537.7 mg, *Hyoscyamus niger* 107.14 mg, Jatamansi root 53.57 mg and *Cannabis sativa* leaves 57.57 mg in a pill (dose 750 mg to 1.25 g) for insomnia. If a patient takes a pill containing 1,125 mg, he will be consuming 178.56 mg of *Hyoscyamus niger* (Paarsika-yavaani). What will be the synergistic action of Paarsika-yavaani with 3 other sedative drugs? It has been recorded that *Hyoscyamus niger* interacts with tricyclic antidepressants, amantadine, anti histamines, phenothiazines, procainamide, and quinidine with concomitant use. (C. P. Khare.)

Ricinus communis Linn.

Eranda, Castor Seed

Toxicity of Beans, Seeds and Oil

Castor seed contains toxic constituents (2.8-3 per cent on whole seed; about 10 per cent in the flour) albumin, ricin, and alkaloid ricinine. An ulcerative factor is also reported. Ingested castor beans are generally toxic only if the ricin is released through mastication. There are no literature reports of poisoning from ingesting purified ricin.

All clinical reports with regard to poisoning refer to castor bean ingestion. Documented mild to lethal clinical symptoms may result from ingesting one half to 30 beans. Symptom onset after ingestion is usually within 4 to 6 hours but may be as late as 10 hours. Initial symptoms are nonspecific and may include colicky abdominal pain, vomiting, diarrhoea, heartburn, and oropharyngeal pain. Hematemesis and melena are reported less commonly. Fluid losses, in some cases may lead to electrolyte imbalances, dehydration, hypotension, and even circulatory collapse.

Concomitant use of castor oil with cardiac glycosides, antiarrhythmic dugs, diuretics, cortisol, liquorice, antihistamines and fat-soluble vitamins may reduce the efficacy of these drugs or, in case of cardiac glycosides, may increase the risk of adverse events due to fluid and electrolyte losses. Due to stimulation of bile flow, the oil should not be used by patients with biliary tract obstruction or biliary disorders, also in cases of intestinal blockage and ileus. The oil is contraindicated in breastfeeding mothers. Ricinoleic acid is absorbed into blood stream and is excreted into human breast milk. In suckling infants it has a purgative effect. (*WHO monographs on selected medicinal plants*, Vol. 4: 280.)

Rubia cordifolia Linn.

Manjishthaa, Indian Madder

Toxicity of Hydroxyanthraquinones

Several substituted napthoquinones and hydroxyanthraquinones and their glycosides have been isolated from the roots of *Rubia cordifolia*. Lucidin primeveroside is also reported. Of these the napthoquinone showed cytotoxic and antitumour activities, while lucidin or aglycons of hydroxyanthraquinones are suspected to have carcinogenic risk in humans.

A group of 30 male ACI/N rats, 1.5 months of age, was fed 1 per cent 1-hydroxyanthraquinone in CE-2 diet throughout the experimental period of 480 days. Thirty control rats were fed basal diet. Rats that survived more than 280 days developed various tumours of the intestine (25/29.) These comprised caecal adenomas (10/29) or adenocarcinomas (5/29) and colonic adenomas (12/29) or adenocarcinomas (11/29.) No such tumours were diagnosed in control rats. In addition, neoplastic nodules and hepatocellular adenomas (12/29) and forestomach papillomas and glandular stomach adenomas (5/29) were observed. No such tumours were observed in control animals. (Mori *et al., Carcinogenesis*, 1990, 11: 799-802.)

A group of 27 male ACI/N rats, six weeks of age, was fed 1.5 per cent 1-hydroxyanthraquinone in the diet for 48 weeks. A second group (14 rats) was also given 16 mg/L indomethacin in the drinking-water for the experimental period. Fifteen control rats were fed basal diet. Rats fed with 1-hydroxyanthraquinone had incidences of 12/27 large intestinal neoplasms (adenomas and adenocarcinomas) and 14/27 forestomach tumours (papillomas.) In rats fed 1-hydroxy-anthraquinone and treated with indomethacin, these incidences were 0/14 and 2/14, respectively. Untreated animals and rats given indomethacin alone had no neoplasms in the large intestine and forestomach (Tanaka *et al., Carcinogenesis*, 1991;12: 1949-1952.)

Calcium Channel Blocking Effect

Crude extract of *Rubia cordifolia* (RC) was tested in isolated tissue preparations for its possible calcium channel antagonistic activity. RC suppressed the spontaneous contractions of guinea-pig atria, rabbit jejunum and rat uterus in a concentration dependent manner (0.1-3 mg/ml.) In rabbit aorta, it inhibited norepinephrine (10 microM) and KCl (80 mM) induced contractions. Replacement of physiological salt solution with calcium free solution abolished the spontaneous movements of rabbit jejunum. However, addition of calcium (25 micrograms/ml) in the tissue bath restored the spontaneous movements. When the tissues were pretreated with plant extract (1 mg/ml) or verapamil (0.5 microgram/ml), addition of calcium failed to restore spontaneous contractions. These results indicate that the plant extract exhibits spasmolytic activity similar to that of verapamil suggestive of presence of calcium channel blocker like constituent(s) in this plant. (HR Ahmed *et al., Journal of the Pakistan Medical Association*, 1994, 44: 82.)

Salacia reticulata Wight

Saptarangi, Salacia

Effect of Root Extract on Reproductive Outcome

The root extract of *Salacia reticulata* is used in Sri Lanka by traditional practitioners as a herbal therapy for glycemic control even during pregnancy. In a study, the effect of root extract on the reproductive outcome of Wistar rats (250-260 g) was determined. The root extract was administered orally (10 g/kg) during early (days 1-7) and mid- (days 7-14) pregnancy. The root extract significantly (P<0.05) enhanced post-implantation losses (control *vs* treatment: early pregnancy, 4.7 ± 2.4 *vs* 49.3 ± 13 per cent; mid-pregnancy, 4.7 ± 2.4 *vs* 41.7 ± 16.1 per cent.) Gestational length was unaltered but the pups born had a low birth weight (P<0.05) (early pregnancy, 6.8 ± 0.1 *vs* 5.3 ± 0.1 g; mid-pregnancy, 6.8 ± 0.1 *vs* 5.0 ± 0.1 g) and low birth index (P<0.05) (early pregnancy, 95.2 ± 2.4 *vs* 50.7 ± 12.9 per cent; mid-pregnancy, 95.2 ± 2.4 *vs* 58.3 ± 16.1 per cent), foetal survival ratio (P<0.05) (early pregnancy, 95.2 ± 2.4 *vs* 50.7 ± 12.9; mid-pregnancy, 95.2 ± 2.4 *vs* 58.3 ± 16.1), and viability index (P<0.05) (early pregnancy, 94.9 ± 2.6 *vs* 49.5 ± 12.5 per cent; mid-pregnancy, 94.9 ± 2.6 *vs* 57.1 ± 16.1 per cent.) However, the root extract appeared to be non-teratogenic. On the other hand, allopathic antidiabetic agents like sulfonylureas and biguanides, and plant products such as *Momordica charantia* fruit extracts induce teratogenesis in rodents.

If these data can be applied to women, then consumption of the root extract during pregnancy can have serious implications in countries like Sri Lanka, India and Nepal where more than two thirds of all infants born are small for gestational age. Low birth weight has a major influence on neonatal morbidity, neurocognitive deficiencies, neurobehavioral effects and mortality. Furthermore, reduced growth *in utero* is reported to be linked to decreased glucose tolerance in adult life. It advised that the use of the *S. reticulata* extract should be avoided by women with pregnancy complicated by diabetes. (W.D. Ratnasooriya *et al., Braz J Med Biol Res*, July 2003, 36(7): 931-935.)

Saraca asoca (Roxb.) De Wilde

Ashoka, Asoka Tree

Estrogenic Activity of Saraca asoca Only in Functional Ovary

In a study, the estrogenic effect of a herbal uterine herbal drug was investigated using *in vivo* and *in vitro* experimental models. The drug contains *Saraca asoca* (10 per cent), *Symplocos racemosa* (6.6 per cent), *Asparagus racemosus* (6.4 per cent) as its main constituents.) Estrogenic effect of the drug (1 gm/kg p.o.) was studied in normal and ovariectomised rats. The parameters studied in both *in vivo* models included uterine weight (wet and dry weight), estrogen and progesterone levels. The effect of the drug (1 g/kg p.o. x 21 days) was also studied in normal rats on regular oestrus cycle. *In vitro* studies with the drug (50-400 µg/ml of aqueous extract) on isolated uterus in non-gravid, non-oestrinised rats were carried out to find out whether the formulation possesses any oxytocin like activity. Administration of the drug in normal rats significantly increased the wet and dry uterine weight. It also resulted in marked increase of estrogen levels with no change in progesterone levels as compared to control. The drug in ovariectomised rats did not show any increase in uterine weight as compared to ovariectomised control. Thus, the drug showed estrogenic activity only in the presence of functional ovary and is devoid of any progestational activity. (Mitra SK *et al., Indian Journal of Pharmacology*, 1999, 31: 200-203.)

Polyalthia longifolia Benth. and Hook. f., an ornamental roadside tree, is wrongly called Ashoka. Its bark is a common adulterant of Ayurvedic Ashoka.

Sesbania sesban (Linn.) Merrill

Jayanti, Common sesban

Toxicity of *Sesbania sesban* Phenolic Compounds

S. sesban does not contain condensed tannins, but does contain phenolic compounds including saponin which has spermicidal and haemolytic activity and was found to depress feeding activity in moth larvae. The inclusion of *S. sesban* in poultry diets (10 per cent of diets) proved fatal to young chicks.

S. sesban ILCA 1198 caused negative effects on oestrus in ewes when fed as a supplement at 13.3 g/kg DM (dry matter) liveweight, reducing the number of ewes showing oestrus by 30 per cent of potential. In comparison, supplementation at 8.2 and 10.9g/kg DM liveweight had no effect on oestrus in ewes. Other studies have reported reproductive irregularities when feeding high percentages (>30 per cent) of *S. sesban* to ruminants. (http://www.tropicalforages.info)

A saline extract of the leaves has been reported to exhibit haemolytic activity. (Dorsaz *et al., Planta Medica*, 1988, 54: 225. *Chem Abstr*, 199, 112, 234102; 1992, 117, 208886.)

The plant contains a rare kaempferol trisaccharide which acts as anti-tumor promotor. (el-Sayed NH, Pharmazie, 1991 Sep, 46(9): 679-80. Cited in *The Wealth of India*, First Supplement Series, Vol. 5, 2004: 82.)

Antidiabetic Activity of Leaf Extract

In an experimental study, aqueous extract of *Sesbania sesban* showed significant increase in serum insulin level. A marked decrease in triglycerides, total cholesterol, LDL and VLDL was observed, while increase in HDL cholesterol has been observed in aqueous leaves extract treated diabetic rats, which suggest that HDL is inversely related to the total body cholesterol. The possible mechanism of antidiabetic action of aqueous extract may be by increasing the pancreatic secretion of insulin from the existing beta cells, by its release from the bound form.

Animals treated with aqueous extract indicated a significant decrease in the glycosylated hemoglobin level which could be due to an improvement in insulin secretion, whereas glycosylated hemoglobin level increased significant in untreated diabetic control group, which confirm the antidiabetic action of the extract (22.) The significant increase was observed in glycogen levels of the aqueous leaves extract treated diabetic rats. The extract did not produce any significant effects on normal animals. (Ramdas B. Pandhare *et al., Avicenna J Med Biotechnol*, 2011 Jan-Mar, 3(1): 37-43.)

Spermicidal Activity of Root Extract

In an experimental study, the spermicidal activity of oleanolic acid 3-β-D-glucuronide (OAG), isolated from root extracts of *Sesbania sesban*, was evaluated by using highly motile rat sperm.

The minimum effective concentration of oleanolic acid 3-β-D-glucuronide was 50 mcg/mL. More than 97 per cent of the OAG-treated sperm lost their hypo-osmotic swelling (HOS) responsiveness in a dose-dependent manner. Transmission electron microscopy (TEM) and sperm membrane lipid peroxidation (LPO) revealed that OAG affected the sperm membrane integrity. OAG declined fertility to zero, was nonmutagenic and was not harmful to lactobacillus. (Nilanjana Das *et al.*, *Contraception*, February 2011, 83(2): 95-188. *ScienceDirect.* Elsevier.)

Semecarpus anacardium Linn.

Bhallatak, Marking-Nut

Bhallatak is classified in Ayurveda under the category of toxic plants. The drug is avoided in paediatric age group, pregnant women, and also in certain diseased conditions such as bleeding diatheses, renal function disorder, history of vesications. It is known to have a narrow therapeutic range.

Determining the Dose for Cancer Chemotherapy

A toxicological study was carried out in rats with chloroform-soluble fraction of the nuts of *Semecarpus anacardium* to determine its safe non-toxic dose. The fraction produced toxicity at all levels tested (50-400 mg/kg) but the extent of toxicity was found dose-dependent. At lower doses this fraction induced partial growth inhibition over 36 days and higher doses proved fatal within 6 days. It was observed that 230 mg/kg caused 50 per cent mortality in rats and this value is 1380 mg/m^2 when expressed for body surface area. These findings will be of some use in the cancer chemotherapy study of the fraction. (Kesava Rao KV *et al., Indian J Physiol Pharmacol*, 1979 Apr-Jun, 23(2): 115-20.) PMID: 489092.)

Cytotoxicty of the Nut Extract

In a study, *Semecarpus anacardium* nut extract was examined for inhibitory effect on human breast cancer cells (T47D.) Cytotoxicity analyses suggested that these cells had become apoptotic. *Semecarpus anacardium* was discovered to induce rapid Ca^{2+} mobilization from intracellular stores of T47D cell line, and its cytotoxicity against T47D was well correlated with altered mitochondrial transmembrane potential. At the molecular level, these changes are accompanied by decrease in Bcl(2) and increase in Bax, cytochrome c, caspases and PARP cleavage, and ultimately by internucleosomal DNA fragmentation. Taken together, these results provide unprecedented evidence that SA triggers apoptotic signals in T47D cells. (Mathivadhani P *et al., Cell Biol Int*, 2007, 31: 1198–206.)

Solanum xanthocarpum Schrad. & Wendl.

Kantakaari, Wild Eggplant

Toxicity on Cell Membrane

Administration of α-solanine (75-100 mg/kg body weight) on a daily basis found to be lethal in hamsters within 4-5 days. Solanine treated animals were also suffered from other undesired effects such as fluid-filled and dilated small intestines. α-solanine induced craniofacial malformations (exencephaly, anophthalmia and encephalocele) on oral administration in Syrian hamsters. These toxic effects of solanine were attributed by the ionic imbalance in the cells. Several studies, on mouse neuroblastoma x rat glioma hybrid NG 108-15 cells, mouse-skin fibroblastoma L-929 cells and mouse Balb/3T3 cells lines, revealed that the solanine-evoked Ca^{2+} influx due to the destabilization of the cell membrane. Further, all the solanine treated cell lines showed a marked increase in intracellular Ca^{2+} concentrations with the concentration of solanine. (Jayakumar K and Murugan, *Journal of Analytical and Pharmaceutical Research*, 2016,03: 00075. DOI: 10.15406.)

Spermatogenesis

Inhibition of spermatogenesis and testosterone production and reduced movement of sperm was found in solasodine treated groups, without any remarkable change in size of sex organs. Above optimal doses, anabasine is believed to be teratogenic in swine. (Jayakumar K and Murugan, *Journal of Analytical and Pharmaceutical Research*, 2016,03.)

Neuro- and Gastrointestinal Toxicity above Optimal Doses

Many of the *Solanum* alkaloids are shown to have regulatory activity on nervous system at low dosage regimens, but exert its neurotoxicity above optimal doses. Overdoses could lead to gastrointestinal toxicity. At a dose of 200-400 mg solanine can cause tachycardia, dyspnea, vertigo, twitches in arms and legs, and cramping. Green fruits have higher solanine concentrations. (Ji YB, Gao, Study on mitochondrion pathway of the apoptosis of HepG2 induced by solanine, *J Chin Pharmaceut Sci, 2008*, 43(4): 272-275. Aslam RP, Gautam PV, Antifertility effects of solasodine obtained from *Solanum xanthocarpum* berries in male rats and dogs, *Int J Pharm Tech 2013*, 4(4): 2305-2310.)

Sphaeranthus indicus Linn.

Munditika, Mundi

Cytotoxic Potential on Human Cell Lines

The study was designed to screen *Sphaeranthus indicus, Ganoderma lucidum,* and *Urtica dioica* for their anticancer activity against human cancer cell lines. Petroleum ether, ethanolic, and aqueous extracts of *S. indicus, G. lucidum,* and *U. dioica* were subjected to cytotoxicity studies using 7 different cancer cell lines. Potent cytotoxicity was noted in petroleum ether extract of *S. indicus* (SIP), which inhibited proliferation of various cancer cell lines. Growth inhibition was determined by sulforhodamine B assay. Two biochemical markers, beta-sitosterol and 7-hydroxyfrullanolide, were isolated and characterized using high-performance thin layer chromatography, melting point, Fourier transform infrared spectroscopy, nuclear magnetic resonance spectroscopy, and mass analysis. Cytotoxicity of isolated beta-sitosterol and 7-hydroxyfrullanolide were also determined. The IC(50) of the extracts of *S. indicus* was calculated in the HL-60 cells and it was found to be 53 μg/mL. Furthermore, SIP induced apoptosis in human leukemia HL-60 cells as measured by several biological end points. Cell cycle analysis and change in mitochondrial membrane potential was quantified by flow cytometry. Subsequently, using annexin V/PI assay, proportion of cells actively undergoing apoptosis was determined. Changes in DNA were observed by DNA ladder assay. The extracts of *S. indicus* induced apoptotic bodies formation, induced DNA laddering, enhanced annexin-V-FITC binding of the cells, increased sub-G(O) DNA fraction, and induced loss of mitochondrial membrane potential ($\Delta\Psi$m) in HL-60 cells. SIP also elevated the caspase 3 and caspase 9 levels in the HL-60 cells, which clearly indicates the involvement of the intrinsic proteins in inducing apoptosis.

All the above parameters revealed that *S. indicus* induced apoptosis through the mitochondrial-dependent pathway in HL-60 cells. The criterion for anticancer activity in cytotoxicity assay was ≥70 per cent growth inhibition at 100 μg/mL against at least 4 cell lines. As *G. lucidum* and *U. dioica* did not exhibit appreciable inhibitory activity against human cancer cell lines (less than 50 per cent), they were not included in the study thereafter. The results established that SIP has apoptosis-inducing effect against HL-60 cells in vitro and is a promising candidate for further anticancer study. β-Sitosterol and 7-hydroxyfrullanolide can be considered to be potent anticancer compounds isolated from the extracts of *S. indicus* on the basis of present studies. (Nahata A *et al., Integr Cancer Ther,* 2013 May, 12(3): 236-47. Epub 2012 Aug 22.)

Potential use in BPH

Effect of *Sphaeranthus indicus* extracts on prostatic hyperplasia induced by testosterone in albino rats was assessed by *in vitro* studies. From the petroleum ether, ethanolic and aqueous extracts of *S. indicus*, the biochemical marker, beta-sitosterol, was isolated and extracts were characterized utilizing HPTLC. Testosterone (3 mg/kg s.c.) was administered to the rats along with the test extracts and isolated beta-sitosterol for a period of 28 days. The weight of the rats, the urine output, serum testosterone concentrations and prostate-specific antigen (PSA) levels were recorded. The prostate/body weight ratio was calculated and histological studies were performed to observe the changes in the histoarchitecture of the prostate. Finasteride was used as a positive control (1mg/kg p.o.) *S. indicus* extracts attenuated the increase in the P/BW ratio induced by testosterone in the treated groups. The petroleum ether extract exhibited the best activity, although the ethanol and aqueous extracts also exhibited significant activity. Urine output was also improved significantly, demonstrating the clinical implications of the study. Histological studies, testosterone levels which were measured weekly and PSA levels measured at the end of the study also support claims for the potential use of *S. indicus* in the treatment of prostatic hyperplasia. (Nahata A, Dixit VK, *Phytother Res*, 2011 Dec, 25(12): 1839-1848. Epub 2011 Apr 19.)

Streblus asper Lour.

Shaakhotaka, Sandpaper Mulberry

Biological Activity of Root bark

The total ethanolic extract of the root bark of *Streblus asper* was found to indicate significant activity on isolated frog heart, blood pressure, isolated rabbit intestine and guinea pig uterus. A alpha-beta-unsaturated lactone was separated. It administered by *i.v.* route gave the LD_{50} of 4.8 mg/kg in white mice. Examinations on isolated frog heart indicated that it induces a stimulating ionotropic effect in 10^{-5} dilution and a systolic response in 10^{-4} dilution. Pronounced *in vitro* spasmodic effect of the compound was observed on the smooth muscles of the rabbit intestine and guinea pig uterus in those high dilutions. (Subha Rastogi, *Evid Based Complement Alternat Med,* 2006 Jun, 3(2): 217–222.)

Antifilarial Compounds

The active antifilarial compounds are cardiac glycosides, they can produce cardiotoxicity and thus it is required to separate the two activities. Attempts in this direction have been made by subjecting strebloside and asperoside to hydrogenation to reduce the alpha-beta- unsaturated lactone ring. The results depicted that at a dose of 50 mg/kg orally, even though there was a decrease in the macrofilaricidal activity showed by dihydroasperoside as well as dihydrostrebloside, but there was an obvious absence of cardiotonic potential as compared to the parent compounds. (Rastogi S, Chemical investigation of biologically active plants *viz., Streblus asper, Bacopa monniera, Amoora rohituka, Bergenia stracheyi* and *Mallotus nepalensis,* PhD Thesis, *Central Drug Research Institute,* Lucknow, India. 1994: 189. Cited in *International Journal of Research and Development in Pharmacy and Life Sciences (IJRDPL),* December-January 2017, 6(7): 2845-2849.)

The glycoside (K030), isolated from the plant in The Central Drug Research Institute, Lucknow, India, caused death of bovine filarial parasite, *Setaria cervi,* within 2-3 hours at concentrations of 10 mg/ml (1.7 pM.)

Strychnos nux-vomica Linn.

Vishamushti, Nux-vomica

Effects of Modified Alkaloid Fraction

A study was carried out to investigate the effect of the total alkaloid fraction extracted from *Nux vomica* on analgesic and anti-inflammatory activity and pharmacokinetics after transdermal administration.

In contrast to the total alkaloid fraction, most strychnine was removed from modified total alkaloid fraction and the ratio of brucine to strychnine was adjusted from 1: 1.8 to 2.7: 1. Modified total alkaloid fraction showed significant analgesic activity in all the chemical-, thermal- and physical- induced nociception models, which indicated the presence of both centrally and peripherally mediated activities. Modified total alkaloid fraction also showed significant anti-inflammatory activity against xylene-induced ear edema. But total alkaloid fraction and strychnine demonstrated little activity in all those pharmacological tests. Brucine showed to be effective in acetic acid-induced writhing and xylene-induced ear edema test. Brucine in modified total alkaloid fraction was absorbed more completely than it alone at the same dosage after transdermal administration.

The results from the study appeared to support the viewpoint that most strychnine should be removed from total alkaloid fraction to improve analgesic and anti-inflammatory activity. The relatively higher pharmacological activity of modified total alkaloid fraction compared to brucine alone is partly due to the enhanced transdermal absorption of brucine. (Jun Chen *et al., Journal of Ethnopharmacology*, 2012 Jan 6, Vol. 139 (1): 181-188.)

Toxicity of Raw and Detoxified Seed

In a study, four different samples of *Nux vomica* seeds and doses were used for the toxicity evaluation in Wister rats. Sample I: Raw, unprocessed, non-shodhit) seeds, 2000, 300, 50, 15 mg/kg. Sample II: Processed, cow's urine- milk- and ghee-shodhit, 600, 300 mg/kg. Sample III: Processed, Kanji-shodhit, 300, 50 mg/kg. Sample IV: Processed, *Aloe vera*-shodhit, 300, 50 mg/kg. (Shodhit=detoxified.)

Sample I was administered to 5 males and 5 females rats. All the rats died within 15 minutes of exposure. Hence a lower dose with 300 mg/kg body weight was considered with only 3 females per dose following the OECD Toxic Class Method, wherein all the animals died within 45 minutes. A next lower dose of 50 mg/kg body weight with 3 females also resulted in lethality within 1 h. There was no mortality in 15 mg/kg body weight dose. This was reconfirmed with another set of females (confirmatory test) treated with 15 mg/kg body weight group treated with Sample I. It was established that the LD_{50} range for Sample I was between 5-50 mg/kg body weight.

Based on data from Sample I, next three groups of animals were exposed to a dose of 300 mg/kg body of Sample II, Sample III and Sample IV. While there was no lethality in the group of animals of Sample II, lethality was established in Sample III and Sample IV at 300 mg/kg body weight dose. For Sample II the lethality was established at 600 mg/kg body weight and for both Sample III and Sample IV no lethality was observed at a lower dose of 50 mg/kg body weight. The LD_{50} range for Sample I was between 5-50 mg/kg body weight, for Sample II between 300-600 mg/kg body weight, (VGS Sharma, MN Reddy, *Int. J. Pharm. Sci. Rev. Res*, September-October 2016, 40(2), Article No. 09: 33-35.)

Symplocos racemosa Roxb.
Lodhra, Lodh Tree

Experimentally, Decreased Testosterone Levels in PCOS

In a study, the anti-androgenic properties of *Symplocos racemosa* in the treatment of hyperandrogenemia associated polycystic ovary syndrome (PCOS) were investigated in a letrozole induced PCOS rat model. The testosterone levels were used to evaluate the anti-androgenic effect of *S. racemosa* in letrozole-induced PCOS rats for 21 d. The low (250 mg/kg), mid (500 mg/kg) and high dose (1000 mg/kg) of *S. racemosa* was given to the PCOS induced rats for 15 d post letrozole induction to determine the effective dose of *S. racemosa* in the treatment of hyperandrogenemia associated PCOS. The hormones such as estrogen and progesterone were also assayed along with testosterone to determine the fluctuations in sex steroid levels in PCOS rats induced by letrozole. *S. racemosa* treatment significantly decreased testosterone levels which were found to be elevated in PCOS rats induced by letrozole. *S. racemosa* significantly restored other blood biochemical parameters such as estrogen, progesterone and cholesterol levels. It also restored the histology of ovarian tissue. The ovarian weights and uterine weights were also significantly recovered after the *S. racemosa* treatment. The mid dose (500 mg/kg) and high dose (1000 mg/kg) of *S. racemosa* were found to be effective in the treatment of hyperandrogenemia in PCOS. This effect of *S. racemosa* was found to be comparable with clomiphene citrate, (the major medicine used in the treatment of PCOS. (Mamta Jadhava *et al., Journal of Coastal Life*, 2013, 1(4): 309-314. ResearchGate synopsis.)

Effect of Anti-androgenic properties on Male Hormones

Antiandrogens, also known as androgen antagonists or testosterone blockers, are a class of drugs that prevent androgens like testosterone and dihydrotestosterone (DHT) from mediating their biological effects in the body. They act by blocking the androgen receptor (AR) and/or inhibiting or suppressing androgen production. (*Mowszowicz I, Antiandrogens: Mechanisms and paradoxical effects, Ann. Endocrinol, Paris*, 1989, 50(3): 189-199.)

In males, the major side effects of antiandrogens are de-masculinization and feminization. These side effects include breast development/enlargement, reduced body hair growth/density, decreased muscle mass and strength, feminine changes and reduced penile length and testicular size. In addition, antiandrogens can cause infertility, osteoporosis, sexual dysfunction (including loss of libido and erectile dysfunction), and decreased semen/ejaculate volume in males. (*Higano CS, Side effects of androgen deprivation therapy, Urology*, 2003, 61: 32–38.)

Caution for *Asavas* and *Arishtas*

Rodhraasava (*Lodhraasava*) is an OTC product. Ayurvedic *Aasava* and *Arishta*, which contain self-generated alcohol, are available without a prescription of a qualified Ayurvedic *vaidya*.

Recently, Cyriac Abby Philips, a liver specialist at the liver unit of Cochin Gastroenterology Group, reported a case of an Ayurvedic tonic-induced severe liver disease in a farmer who never drank alcohol. The patient was a 40-year-old man who came to the out-patient department for evaluation of severe jaundice. His bilirubin levels were in excess of 12 mg/dl and his liver enzymes were more than five to eight times the upper limit of normal. His liver tests had all the indications of alcohol use, but the man denied having taken a single drop of alcohol all his life.

He was a pineapple farmer, led a healthy life and did not indulge in any substance abuse. He was fit and without a shred of excess fat. Every common cause was searched for viruses, autoimmune hepatitis, use of over the counter medications, herbal medications, painkillers, antibiotics. All came up as negative. This was first time the farmer had ever fallen sick. Then the patient was investigated for rarer causes like herpes viruses, cytomegalovirus, parvovirus, dengue, typhoid fever and rare cancers. All these blood tests came back negative too. The liver biopsy was done. It revealed severe alcoholic hepatitis. Alcoholic hepatitis strikes the liver of a person who drinks heavily regularly or who binge drinks. But the patient never took even a single drop of alcohol in my entire life. All his family members vouched for the patient's version—he never drank. But he eats was taking a digestive to reduce the bloating after the meal. But then the man's wife disclosed that he takes a digestive to reduce the bloating after the meal for so many years. It was *Dashamoolarishtam*.

Dashamoolarishtam is an age old Ayurvedic tonic used for supposedly improving digestion and a multitude of other symptoms. It is sometimes the second-most common household item in a Kerala home after the electric mosquito-killer bat. *Dashamoolarishtam* is made up of more than 30 herbal and other ingredients and has self-generated alcohol due to the presence of extracts from the *Woodfordia fruticose* plant that, like grapes, can ferment. *In recent times, some practitioners have added baking yeast to speed up the fermentation process. This increased alcoholic content, which gives the feeling of relief in digestive problems. Some practitioners also add alcohol directly in the tonic.* This Ayurvedic medicine is so popular in Kerala, that when the government banned alcohol sales in the state, people started overdosing on *Arishtams* to get a high. Some arishtam producers intentionally increased alcohol content in their product to boost sales.

The patient was consuming around four to five ounces of *Dashamoolarishtam* four times a day after every meal. That is almost 38 grams of alcohol every day, considering alcohol content in the *Arishta* to be 8 per cent to 10 per cent. This was what caused his severe alcohol-related liver injury.

The pathologist who saw the liver biopsy also reported that a finding not usually seen with alcohol liver injury–necrosis. In the past month, the patient had been drinking a lot of fresh pineapple juice harvested from his own crops in which large amounts of pesticides and insecticides like Fenval and Karate were used. The patient was treated for Ayurvedic *Arishta*-related severe alcoholic hepatitis and chemical-induced toxic hepatitis, He recovered fully after treatment.

According Dr Chandra Kant Katiyar, Chief Executive of one of the most reputed Ayurvedic drug manufacturing company, if we consider minimum 8 per cent alcohol in *Dashmuulaarishta* and some one consumes 38 g it is a very high dose. Normal dose is 15 ml twice daily, *i.e.* 30 ml. At the rate of 8 per cent, it comes to only 2.4 g of alcohol. (*Personal communication.*) No *Vaidya* will disclose the possible toxicity of Ayurvedic drugs and patients go on consuming non-standardized Ayurvedic drugs as and when required.

Syzygium aromaticum (L.) Merr. & Perry

Lavanga, Clove

Blocking Effect on Calcium Channels

Constituents of the clove oil have blocking effect on calcium channels and thereby diminish intercellular calcium concentration. (*Planta Med*, 1993, 59 *Suppl*, A687.)

In another investigation, it was observed that constituents of clove oil (eugenol and beta-caryophyllene oxide) which have relaxant effects in smooth muscle preparations also block calcium channels of the heart. A comparison of the negative inotropic effects of these substances with that of the classic calcium channel blocker, nifedipine, showed that they differ with regard to their calcium channel blocking capacity. Nifedipine exerted the strongest effect on the force of contraction, which was much higher than expected from its calcium channel blocking effect. In contrast, the decrease in the force of contraction after the application of beta-caryophyllene oxide was comparable to that which could be expected from its blocking effect on calcium channels. The results showed that there are calcium channel blocking substances which have a much lower negative inotropic effect on heart muscle than others, which may be explained by the additional influence on other ion channels. (Oliver Sensch *et al., Br J Pharmacol*, 2000 Nov, 131(6): 1089-1096.)

Adverse Effect of Overdose

The clove oil should be used only in restricted concentration. Eugenol in the oil causes hepatotoxicity similar to that associated with paracetamol overdose. (*The Wealth of India, Second Supplement series*, Vol 1.)

Syzygium cuminii (Linn.) Skeels

Jambu, Java Plum

Effect of Jambu Fruit Pulp on Fasting Blood Sugar Levels

In a clinical trial by C.V. Nande, P.M. Kale, S.Y. Wagh, D.S. Antarkar (R.R.A. Ayurvedic Research Institute, Mumbai) and A B Vaidya (Former Director, Ciba-Geigy Research Centre, Mumbai) the fruit pulp of 150 g Jambu was investigated in 6 nondiabetics and 5 diabetic subjects. In nondiabetics (average pulp ingested 115.2 g) average fasting blood sugar 75.3 dropped to 68.1 after 1 hour, 67.1 after 2 hours and 68.3 after 3 hours. In 5 diabetics (average pulp ingested 112.1 g) average fasting blood sugar 240 was found elevated to 254.4 after 1 hour, 256.6 after 2 hours and 272.6 after 3 hours. (*Journal of Resarch in Ayurveda and Siddha*, Vol. IV (1-4), 1-5.)

CNS Depressant Activity of Seeds

Neuropsychopharmacological studies have been conducted on the chloroform fraction extracted from the seeds of *Syzygium cuminii* L., on experimental animals. The extract was found to produce alteration in general behaviour pattern, reduction in spontaneous motility, hypothermia, potentiation of pentobaritone hypnosis, analgesia, reduction in exploratory behaviour pattern, muscle relaxant action, and suppression of aggressive behaviour. The extract also caused suppression of conditioned avoidance response and showed antagonism to amphetamine group toxicity. These observations suggest that the extract of the seeds of *S. cuminii* possesses a potent CNS depressant action. (D Chakraborty *et al., Planta Medica*, May 1986, 52(2): 139-143.)

Mixed Results of Hypoglycemic Activity

Clinical and animal research indicate the Jambu leaves and fruit do not have antidiabetic activity. Jambu seed or bark also has not demonstrated hypoglycemic or hypolipidemic effect in humans. (*Natural Medicines Comprehensive Database*, 2013: 920.)

Terminalia arjuna (Roxb.) Weight &Arn.

Arjuna

Anti-implantation, Abortifacient, Absorptive Activity of the Bark

The bark is cardioprotective, but in an experimental study the crude bark and ethanolic extract of the bark, when administrated to pregnant rats, showed significant anti-implantation and abortifacient, absorptive activity, but did not influence spermatogenesis. The crude drug showed anti-implantation as well as foetus absorption activity, while ethanolic extract showed anti-implantation activity only. (Lal and Udupa, *J Res Ayurv Siddha*, 1993, 14, 165; Karamsetty *et al., Phytother Res*, 1995, 9, 575; *Chem Abstr*, 1995, 123, 25229.) The stem bark exhibited oxytocic activity, attributed to arjunolone. (Rastogi and Dhawan, *Indian J Med Res*, 1982, 76 *Suppl.*)

Effects of Bark Extract on Gonadal Steroidogenesis

One experimental study evaluated the effects of bark extract on gonadal steroidogenesis, spermatogenesis and plasma gonadotrophin (FSH and LH) levels in male albino rats.

Adult rats were administered orally with alcoholic extract of its bark dissolved in distilled water at the dose of 150 mg kg^{-1} day^{-1} for 7, 14 and 21 days. Testicular steroidogenesis was evaluated by spectrophotometric assay of the activities of 17beta-HSD and delta^5-3beta-HSD and radioimmunoassay of plasma testosterone level. Plasma FSH and LH levels were also determined by radioimmunoassay. Spermatogenic activity was evaluated by counting the germ cells at stage VII of seminiferous cycle. Treatment for 21 days led to significant decrease in the activities of steroidogenic enzymes, plasma testosterone level and gametogenic activity. These parameters showed non-significant changes in the 7 and 14 days treatment group and no significant change was found in plasma gonadotropin levels at any of the above durations. Though the exact mechanism of this antigonadal effect could not be determined from this study, long term adverse effect of *T. arjuna* bark may have some clinical implication. (Alok Chattopadhyay, Ronojoy Sengupta and Shaba Parveen, Antigonadal Effects of *Terminalia arjuna* in male albino rats. *Research Journal of Medicinal Plants*, 2014, 8: 82-91.)

Alcoholic extract of the bark of *T. arjuna* fed orally to albino rats at different doses caused degenerative changes in the testis, epithelial atrophy, narrowing and distortion of the lumen and infiltration of lamina propria with wide channels in the vas deferens of rats. (Chauhan S, Agarwal S and Mathur, Vasal assault due to *Terminalia arjuna* bark in albino rats./, 1990, 22: 491-494.)

Terminalia chebula Retz.
Haritaki, Chebulic Myrobalans

Hepatic and Renal Adverse Effects

Dietary administration of the fruit to rats, as 25 per cent of the diet, produced hepatic lesions which included centrilobular vein abnormalities and centrilobular sinusoidal congestion. Marked renal leision were also observed, and included marked tubular degeneration, tubular casts and intertubular congestion. A brown pigmentation of the tail and limbs was also observed after 10 days. The median lethal dose of a 50 per cent ethnol extract of the fruit was 175.0 mg/kg bw after interperitoneal administration. (Abraham Z *et al.*, Screening of Indian plants for biological activity, Part XII, *Indian journal of Experimental Biology*, 1986, 24: 48-68. Cited in *WHO monographs on selected medicinal plants*, Vol. 4, 2009: 78.)

Significant Anti-spermatogenic Effect in Male Rat

In an experiment, aqueous-ethanolic (1: 1) extract of fruit of *T. chebula* was administered orally at a dose of 60 mg/0.5 mL distilled water/day for 28 days. Different parameters were studied including body weight, relative weight of reproductive organ, sperm motility, sperm count, testicular cholesterol, plasma testosterone, testicular androgenic key enzymes, bio-markers of oxidative stress, and histological analysis of the tissues.

Testicular cholesterol showed a significant elevation in *T. chebula* treated group and plasma testosterone was decreased significantly in comparison to control. Anti-oxidative enzymes such as catalase and superoxide dismutase showed a significant reduction and a significant elevation in conjugated diene and thiobarbituric acid reactive substance was noted in treated group. GOT and GPT study of liver and kidney showed a non-significant change which confirmed the non-toxic nature of *T. chebula*. Histological study of testis of treated group exhibited significant reduction in seminiferous tubular diameter. (A. Ghosh *et al.*, *Asian Pacific Journal of Reproduction*, 2015, 4(3): 201-207.)

Trigonella foenum-graecum Linn.
Methi, Fenugreek

Spermicidal Effects

A 50 per cent ethanol extract of the seeds, 2 per cent, *in vitro,* had spermicidal effects and immediately immobilized human sperm on contact. (Dhawan B N *et al.,* Screening of Indian plants for biological activity, Part VI, *Indian Journal of Experimental Biology,* 1977, 15: 208-219.)

Toxicity of Crude Saponins

Administration of a saponin fraction from the seeds by intramuscular injection or by intraperitoneal injection 50.0 mg/kg bw per day, or in drinking-water 500.0 mg/kg bw to chicks for 21 days, decreased bodyweight and increased liver enzymes. Pathological changes included fatty cytoplasmic vacuolation in the liver, necrosis of hepatocytes with lymphocytic infiltration, epithelial degeneration of renal tubules, and resulted in catarrhal enteritis, myositis and peritonitis. (Nakhla HB *et al.,* The effect of *T. foenum-graecum* crude saponins on Hisex-type chicks, *Veterinary and Human Toxicology,* 1991, 33: 561-564.)

No Role of Trigonelline in Hyperglycemia

Administration of isolated trigonelline, in the amounts present in a therapeutic dose of fenugreek, to diabetic patients did not show any significant hypoglycemic activity. (*ESCOP Monographs,* Second Edn: 94.) Antidiabetic and hypocholesterolemic property of the seed has been attributed largely to fenugreek's saponin and high fibre content, and is probably not related to its major alkaloid trigonelline. (Al-Habori M. Raman A, *Phytotherapy Res,* 1998, 12: 233-242.)

Vetiveria zizanioides (Linn.) Nash

Nash Ushira, Cuscus grass

Stimulation of Sympathetic Nerve Activity of the Oil

Although the components of the oil and their biological activities have been studied extensively, the effect of the volatiles emitted from the roots of *V. zizanioides* on humans has so far remained unexplored. In a clinical study, the effects of volatile compounds emitted from the cut roots of *V. zizanioides* (1.0 g, low-dose conditions; 30 g, high-dose conditions) on individuals were investigated. Participants who breathed the volatile compounds emitted under low-dose conditions showed faster reaction times and stimulation of sympathetic nerve activity as measured by electrocardiography. These effects were not observed under high-dose conditions. The total amount of volatiles emitted during the experiment was 0.25 µg under low-dose conditions and 1.35 µg under high-dose conditions.

These findings indicate that volatile compounds emitted from the roots of *V. zizanioides* under low-dose conditions may have helped subjects to maintain performance in visual discrimination tasks while maintaining high sympathetic nerve system activity. (Matsubara E *et al.*, *Biomed Res*, 2012, 33(5): 299-308. PMID: 23124250.)

Effect of Khusimol on Rat Liver

In the course of a random screen of various plant extracts, khusimol, a non-peptide molecule isolated from the root of *Vetiveria zizanioides*, was found to competitively inhibit the binding of vasopressin to rat liver V1a receptors (Ki=50 microM.) The 1H- and 13C-nmr spectra of this sesquiterpene alcohol were assigned unambiguously. (Rao RC *et al.*, *J Nat Prod*,1994 Oct, 57(10): 1329-1335. *PMID*: 7807119.)

Withania ashwagandha Kaul

Ashwagandha

Effect of Ashwagandha on Thyroid Hormones

One Ayurvedic physician, Dr. Ray Sahelian, and The Memorial Sloan-Kettering Cancer Center reported a case where a woman taking *Ashwagandha* for chronic fatigue developed thyrotoxicosis. The condition resolved itself once she stopped using the supplement. The literature did not note dosage or how long she had been taking it. (Ryan Biddulph: Livestrong.com.)

The effects of daily administration of *Withania somnifera* (Ashwagandha) root extract (1.4 g/kg body wt.) and *Bauhinia purpurea* bark extract (2.5 mg/kg body wt.) for 20 days on thyroid function in female mice were investigated. While serum triiodothyronine (T_3) and thyroxine (T_4) concentrations were increased significantly by *Bauhinia*, *Withania* could enhance only serum T_4 concentration. (Panda S, Kar A, *Withania somnifera* and *Bauhinia purpurea* in the regulation of circulating thyroid hormone concentrations in female mice,51, *J Ethnopharmacol*, 1999, 1325-31.)

In a randomized clinical trial in which *Ashwagandha* was used to improve cognitive function in patients with bipolar disorder, laboratory indices of thyroid function (TSH, Free T_4, and T_3) were measured. This was done in light of a case-report of *Ashwagandha*-associated thyrotoxicosis in mice that showed significant increases in thyroxine levels. Ten (of the original 60) patients showed abnormal results in one of the thyroid measures either at the beginning or end of the 8-week study. One *Ashwagandha*-treated patient had subclinical hypothyroidism (TSH, 5.7 mIU/L) at baseline that normalized, and all three *Ashwagandha*-treated patients experienced T_4 increases from baseline (7 per cent, 12 per cent, and 24 per cent.) 6 of 7 placebo-assigned patients showed decreases in T_4 from baseline (4 per cent to 23 per cent), and one patient's TSH moved from the normal to subclinical hypothyroid range (6.96 mIU/L.) As thyroid indices were done for safety, and not the primary goal of the original study, only 16.7 per cent had abnormal thyroid indices, and as there was no sub-stratification for treatment assignment by thyroid status, unequal numbers of subjects received *Ashwagandha* (n =3) or placebo (n=7.) In spite of these limitations, the subtle laboratory changes noted in thyroid indices in an 8-week study suggest that *Ashwagandha* may increase thyroxine levels, and therefore vigilance regarding hyperthyroidism may be warranted. (*J Ayurveda Integr Med*. 2014 Oct-Dec; 5(4): 241–245.)

Marked Impairment in Libido

In a study, male rats were orally administered 3000 mg/kg-day for 7 days. Sexual behaviour of rats was evaluated 7 days prior to the treatment, day 3 and 7 of the treatment, and day 7, 14 and 30 post-treatment by pairing each male with a receptive female. The root extract induced a marked impairment in libido, sexual performance, sexual vigour, and penile erectile dysfunction. These effects were partly reversible on cessation of treatment. These antimasculine effects were not due to changes in testosterone levels or toxicity but might be attributed to hyperprolactinemic, GABAergic, serotonergic or sedative activities of the extract. (Ilayperuma I *et al., Asian Journal of Andrology*, 01 Dec 2002, 4(4): 295-298.)

Subacute Toxicity

Subacute intraperitoneal toxicity studies at 100 mg/kg/day for 30 days in rats and mice led to the decreased spleen, thymus, and adrenal weights, but no mortality or haematological changes were noted. (Sharada AC *et al., J Pharmacognosy*, 1993, 31 (3), 205-212.)

In mice fed the extract of the entire plant as 25 per cent of the total diet, microscopic lesions in the lung and liver were apparent, and vascular and tubular congestion were noted. (Kulkarni SK, Dhir A, *Prog Neuropsychopharmacol Biol Psychiatry*, 2008, 32 (5), 1093-1105.)

Contraindications

Ashwagandha is contraindicated during pregnancy. It can induce abortion, warns the Sloan-Kettering Memorial Cancer Center. At least one constituent in the herb, nicotine, is a uterine stimulant.

Ashwagandha might also interact with digoxin, when prescribed for the symptoms of congestive heart failure. Taking *Ashwagandha* with barbiturates might increase their sedative effects.

Ashwagandha: An Adopted Common Name for all Chemotypes

Ashwagandha is now an adopted common name of *Withania somnifera* which differs in morphology and chemical constituents based on its chemotypes and sub-species and geographical locations.

At least 3 Chemotypes of *Withania somnifera* have been reported. Chemotype I (Israele) contains withaferin A as a major constituent, minor constituents are withanolides N and O. Chemotype II (Israele): contains withanolide D as major constituent, withanolide G, 27-hydroxywithanolide D, 14 alpha-hydroxywithanolide D and 17 alpha-hydroxy withanolide D are present in trace amounts. Chemotype III (Israele) contains withanolides E-M. Leaves of offspring F_3 (from crossing I and III) contains withanolides Q and R. Withanolides P and S in are also present in the leaves. (Akhtar Hussain *et al.*, Central Institute of Medicinal and Aromatic Plants, Lucknow, India.)

Indian species of *Withania* has been identified as *Withania ashwaganda* sp. *novo* (Bilal Ahamad Mir and Sushma Kaul.)

A study was conducted to determine chemical and morphological variability among five *W. ashwagandha* (WA) populations collected from different regions of India. Variation in the contents of three promising bioactive withanolide, namely withanolide A (WS-1), withanone (WS-2) and withaferin A (WS-3) and morphological characters including plant height, number of shoots, root yield, and leaf biomass per plant were investigated. A considerable degree of variation in these bioactive withanolides and morphological characters was detected among the populations. Plant height and plant biomass were the highest in plants collected from Manasa population followed by Hyderabad. Leaves were found to be the principle organ for WS-3 accumulation while roots mainly accumulate WS-1, suggesting a spatial variation of withanolides. Plants of Manasa population alone showed the presence of WS-2. Withanolide accumulation was also the highest in Manasa population, with 1.312% WS-3 in the leaves and 0.083% WS-1 in roots, suggesting that plants from Manasa (WA02) are elite with regard to the parameters investigated. Further, ontogenetic variation of bioactive withanolides was studied in WA02 at five developmental stages. Withanolide accumulation correlated positively with developmental stages and highest content of these withanolides was found at maturity in both roots and leaves indicating that plants be harvested at maturity stage for maximum economic benefit. These results would offer a suitable alternative for unwise random harvesting that leads to rapid reduction of the existing natural populations of *W. ashwagandha* and for use as potential pharmacological agents. (Bilal Ahmad Mir. Jabnnena Khazr. Khalid Rehman Hakeem, Sushma Kaul, Withanolides array of *Withania ashwagandha* sp. novo populations from India, *Industrial Crops and Products*, August 2014, 59: 9-13. Researchgate publication 262452818.)

Woodfordia fruticosa (L.) Kurz

Dhaataki, Fire-flame Bush

Dried flowers of *Woodfordia* are used in the preparation of Ayurvedic *arishtas* and *aasavas*. In 1982, a yeast strain (*Saccharomyces cerevisiae* was isolated from the flowers which was found capable of producing alcoholic fermentation. (Atal C.K. *et al., Jour Res Ayur Siddha,* 1982, (3 and 4): 193-199.)

The flowers form an important ingredient is *aasavs* for uterine diseases and averting abortion, especially for menorrhagia and leucorrhoea. In *Bhavapraksha,* the flower-buds are included in Yoga-Vandhya (compound to treat sterility) and flowers in *Yoga-garbhpaata* (compound to treat threatened abortion.) In *Gadanigraha,* the woman in season was given blue lotus mixed with Dhaataki flowers and honey in the morning for conception.

Antifertility Activity of Dried Flowers

In a study aimed at preliminary phytochemical investigation also investigated antifertility activity of dried flowers of *Woodfordia fruticosa.* The dried flowers were extracted successively with various solvents and individually with water and aqueous alcohol (50: 50.) The extracts were evaluated for phytochemical studies, including qualitative tests and high performance thin layer chromatography (HPTLC) analysis. Antifertility activity of successive alcoholic, individual aqueous and individual hydroalcoholic extracts was studied in female albino rats. The results revealed that the alcoholic extract showed significant abortifacient activity, whereas aqueous and hydroalcoholic extracts showed moderate activity as compared to the control. Thus, the successive alcoholic extract showed significant abortifacient activity at 100 mg/kg body weight. (Khuslani H. *et al., Indian Journal of Pharmaceutical Science,* 2006. 68(4): 528-529.)

Zingiber officinale Rosc.

Aardraka (fresh), Shunthi (dried), Ginger

Studies in Mutagenicity

A study showed that 6-gingerol and 6-shogaol using column chromatography were mutagenic at 700 microMol in the Hs30 strain of *Escherichia coli*. 6-gingerol was noted to be a potent mutagen whereas 6-shagaol was less mutagenic. (Nakamura H., Yamamoto T., *Mutat Res*, 1983, 122: 87-94.)

Antimutagenicity of Zingerone

Ginger extract and its constituents gingerol, shogaol and zingerone were tested in *Salmonella typhimurium* strains TA 100, TA 98, TA 1535 and TA 1538 in the presence and in absence of S9 mix. It was observed that ginger extract, gingerol and shogaol were mutagenic on metabolic activation in strains TA 100 and TA 1535, but zingerone was non-mutagenic in all the four strains with or without S9 mix. When mutagenicity of gingerol and shogaol was tested in presence of different concentrations of zingerone it was observed that zingerone suppressed mutagenic activity in both the compounds in a dose dependent manner. (Nagabhushan M *et al.*, *Cancer Lett*, 1987 Aug, 36: 221-233.)

Effect on Foetal Development

A study investigated the effect of ginger, a common morning sickness remedy, on foetal development. Pregnant Sprague-Dawley rats were administered, from gestation day 6 to 15; 20 g/L or 50 g/L ginger tea via their drinking water and then sacrificed. No maternal toxicity was observed, however embryonic loss in the treatment groups was double that of the controls (P<0.05.) No gross morphologic malformations were seen in the treated foetuses. Foetuses exposed to ginger tea were found to be significantly heavier than controls, an effect that was greater in female foetuses and was not correlated with increased placental size. Treated foetuses also had more advanced skeletal development as determined by measurement of sternal and metacarpal ossification centres. The results of this study suggested that *in utero* exposure to ginger tea results in increased early embryo loss with increased growth in surviving foetuses. (Wilkinson JM, Effect of ginger tea on the foetal development of Sprague-Dawley rats, *Reprod Toxicol*, 2000, 14(6): 507-512.)

Effect on Spermatogenesis in Rats

In an experiment, wistar male rat (n=30) were allocated into three groups, control (n=10) and test groups (n=20), that subdivided into groups of 2 that received ginger rhizome powder (50 and 100 mg/kg/day) for 20 consequence day. Animals were kept in standard conditions. In twentieth day the testes tissue of rats in whole groups were removed and sperm was collected from epididymis and prepared for analysis. Serum total testosterones significantly increased in experimental group that has received 100 mg/kg/day Ginger (p<0.05) in comparison to control group. Besides, the percentage of sperm viability and motility in both test groups significantly increased (p<0.05) in comparison to control group, Whereas, LH, FSH hormones, sperm concentration, morphology and testes weights in both experimental and control group were similar. (Arash Khaki *et al., Iranian Journal of Reproductive Medicine, Winter* 2009, 7(1): 7-12.)

Fibrinolytic Activity

In a study, Ginger given to patients with coronary disease Ginger (4 g daily for 3 months) did not affect platelet aggregation induced by either adrenaline (epinephrine) or ADP and no changes were observed in fibrinolytic activity or fibrinogen levels. However, a single dose of 10 g of Ginger did produce a significant reduction (p<0.05) in agonist- induced platelet aggregation. (*ESCOP Monographs*, Second Edn, 2003: 551.)

Antimigraine Effects

Mechanism of headache amelioration of Ginger may include its inhibition of thromboxane production and inhibition of free radicals formed in the arachidonic acid cascade. Ginger decreases platelet aggregation and is a potent inhibitor of prostaglandins, which enhance release of substance P from trigeminal fibres (opiates inhibit substance P release.) *(PDR for Herbal Medicines, 2007: 366.)*

Interaction with Anticoagulants

A probable interaction of Ginger with oral anticoagulant phenprocoumon, resulting in over-anticoagulation has been reported in a patient on long term anticoagulation therapy. However, in healthy subjects who received an extract equivalent to 3.6 g of Ginger daily for 7 days, no significant effects were observed on platelet aggregation and coagulation, and after co-administration of a single 25 mg dose of *rac*-warfarin no significant effects of Ginger were apparent. (*ESCOP Monographs*, Second Edn. Supplement, 2009: 297.)

Ginger is found in the official pharmacopoeias of Austria, China, Egypt, Great Britain, India, Japan, the Netherlands, and Switzerland.

Toxicity, Bio-interactions and Herb-Drug Interactions Based on Standard Reference Works

Abies webbiana Lindl.
Taalisa Patra, Fir needle

Abies alba Needle oil: Bronchospasm can be increased.[1] Fir shoots can exacerbate asthma and whooping cough.[1,2b]

Abrus precatorius Linn.
Gunjaa, Jequirity

Toxic constituent: Alkaloid abrine, cytotoxic potential.[4b]

Adverse effects, including death, may occur up to 10-14 days after ingestion of seeds.[2b]

Abutilon indicum (Linn.) Sweet
Atibalaa, Country Mallow

Plant considered as an abortifacient.[4c]

Country Mallow with ephedrine content is banned in the U.S.

Acacia arabica (Lam.) Willd.
Babbula, Indian Gum Arabic Tree

Acacia gum: Contains peroxidase enzyme which is typically destroyed by brief exposure of heat. This enzyme forms coloured complexes with certain amines and phenols and enhances the destruction of many pharmaceutical products, including alkaloids.[3]

When exposed to alcohol, becomes insoluble and will form a precipitate.[5,6] Acacia can reduce the absorption of Amoxicillin.[2a,b]

Contraindicated in intestinal obstruction.[5]

Acacia catechu (L. f.) Willd.
Khadir, Cutch Tree

Leukoagglutinating activity of saline extract of seeds against leukemic cells is inhibited by simple sugars.[4c] Catechin (cianidanol) is associated fatal haemolytic anaemia.

Unstandardized products may contain high amount of aflatoxin, metabolite of Aspergillus, which is toxic and may lead to certain cancers.[2]

Cutch interacts with antihypertensive drugs.[2b]

Acacia concinna (Willd.) DC.
Saptalaa, Soap-nut Acacia

Bark saponins: Spermicidal.[4b]

Acacia leucophloea (Roxb.) Willd.
Irimeda, White Babul

Root: Abortifacient.[4c]

Achyranthes aspera Linn.
Apaamaarga, Prickly Chaff Flower

The benzene extract of the plant exhibited (100 per cent) abortifacient activity experimentally.[4c] Acetone and methanolic extract of the root exhibited anti-implantation activity in rats.[4d]

Aconitum ferox Wall. ex Ser.
Vatsanaabha, Indian Aconite

Aconitine: 2 to 5 mg may cause death. Even external application is reported to cause toxic symptoms.[3] Aconite interacts with Antiarrhythmics, antihypertensives, and Cardiac glycosides. Aconitine is a cardiotoxin that can cause tachyarrhythmias.[5]

Toxicities and fatalities can also occur after ingestion of (Chinese) cured and processed aconite roots.[2b]

Acorus calamus Linn.
Vacha, Sweet Flag

Drugs interacting with the herbs: Antacids, CNS depressants, H_2 Blockers, Monoamine Oxidase Inhibitors, Proton Pump Inhibitors.[2a] Contraindicated during pregnancy.[17(2)]

Actiniopteris dichotoma Kunh
Vahirshikhaa, Peacock's Tail

Ethanolic extract (50 per cent) of the plant showed antispermatogenic activity in male rats.[4d]

Adhatoda vasica Nees.
Vasaka, Malabar Nut

Vasicine and vasicinone may potentiate bronchodilatory activity of theophylline and isoprenaline.[9]

Aegle marmelos (L.) Correa ex Roxb.
Bilva, Bengal Quince

Aqueous and alcoholic extract of leaves are reported to possess effect like digitalis on amphibian and mammalian hearts.[4b]

Aurapten is found comparable with verapamil.[4c]

Marmelosin, like psoralen, causes an increased deposition of pigment melanin by augmenting enzymatic activity.[4d]

Aerva lanata (L.) Juss. ex **Schult.**
Bhadraa, Mountain Knotgrass

(A substitute for Paasaanabheda. AFI.)

Plant extract did not show any significant diuretic activity in healthy human volunteers *(Indian J Physiol Pharmacol,* 1993, 37: 135.)

Ailanthus excelsa Roxb.
Aralu, Tree of Heaven

An alcoholic extract of leaf and stem bark showed anti-implantation and early abortifacient activities in female rats.[4c]

Alangium salviifolium (Linn. f.) **Wang.**
Ankola

Flowers contain deoxytubulosine, a potent antiplatelet aggregation component which has a strong binding with DNA.[4c]

Albizia lebbeck (Linn.) **Willd.**
Shirisha, Siris Tree

Seed saponin: Spermicidal.[4c]

Albizia procera **Benth.**
Katabhi, White Siris

Seed saponin: Spermicidal.[4c]

Allium cepa **Linn.**
Palaandu, Onion

Contraindicated in bleeding disorders, uncontrolled diabetes. Daily maximum amount of diphenylamine: 0.035 g.[1]

Allium sativum **Linn.**
Lashuna, Garlic

Contraindicated in bleeding disorders, gastric ulcer, thyroid disease.[5] Inhibits platelet aggregation, shows additive anticoagulant, antiplatelet effects.[12] Harmful interaction with heparin, Warfarin, NSAIDs.[13]

To be discontinued atleast 7 days prior to surgery.[5] Blood clotting time has been reported to double in patients taking warfarin and garlic supplement.[16a]

Alocasia indica (Lour.) Spach.
Maanaka, Giant Taro

Leaves, stalks, tubers and roots contain high concentration of soluble oxalates. Prolonged use may lead to calcium deficiency and oxaluria.

All parts, except rhizome, contain cyanogenic principle.[4b]

Aloe barabdensis Mill.
Ghritkumari, Aloe

Dried latex: Contraindicated in intestinal obstruction, acutely inflammed intestinal diseases, *e.g.*, Crohn's disease, ulcerative colitis, appendicitis, abdominal pain of unknown origin. Not to be prescribed during pregnancy or to children under 12 years of age.[1,10] Avoid during the first trimester of pregnancy or take under medical supervision.[8b]

Toxic constituents in dried latex: Anthraquinone glycosides.[11] (Not found in *A. vera* gel.)[11]

A. vera latex: Interacts with cardiac glycosides and thiazide diuretics; can cause electrolyte imbalance, may potentiate drug toxicity.[12] Aloe may increase potassium loss in GI, which may be additive with potassium loss due to antiarrhythmics, cardiac glycosides, and corticosteroids.[5] (Sharon M Herr, Herb-Drug Interaction Handbook, 2nd Edn: 21-22.)

Drugs interacting with the herb: Antidiabetic drugs, Digoxin, diuretic drugs, Sevoflurane, stimulant laxatives.[2a] Long term abuse as a laxative may enhance hypokalaemia by thiazide diuretics, adrenocorticosteroids, and licorice root.[16a]

Alstonia scholaris (Linn.) R. Br.
Sapta-parna, Dita Bark Tree

Alkaloid echitamine (from the bark) is found to be toxic to mice in doses of 0.3 - 0.5 mg/20g body wt.[4b]

Alternenthera sessilis (Linn.) R. Br. ex DC.
Matsyaakshi

Alkaloidal extract of the plant in moderate and high dose levels produce slight hepatotoxic and nephrotoxic effects.[4d]

Amanita muscaria Linn.
Soma (one of the main drugs) Fly Agaric (Mushroom)

Toxic constituents: Ibotenic acid and muscimol (isoxazole derivatives), muscazone (oxazole derivative.)

Small amount of muscarine (alkaloid), stizolic acid, tricholomic acid. Isoxazole constituents are psychoactive.[2a,11]

Amaranthus spinosus Linn.
Alpmaarish, Spiny Amaranth

The plant gave negative antibiotic tests, but has a high phagocytic index.[4b]

Amorphophalus campanulatus Blume ex Decne.
Suurana, Elephant-foot Yam

Alcoholic extract of the plant oxytocic.[4d]

Anacyclus pyrethrum DC.
Aakaarakarabha, Pyrethrum Root

In larger doses, the powdered root is an irritant to the mucous membrane of the intestine causing bloody stools, tetanus-like spasms and profound stupor.[4b]

Andrographis paniculata Wall. ex Nees
Kaalmegh, Creat

Contraindicated in bleeding disorders, hypotension, male sterility.[5]

Adverse reactions: GI distress,[5,10] anaphylaxis, infertility.[5]

Patients with autoimmune diseases, including rheumatoid arthritis to avoid or use Andrographis with caution.[2] Herb should not be used during pregnancy and lactation.

Potential antagonism exists between the herb and endogenous progesterone (*Pharmacopoeia of the People's Republic of China*, Vol. 1.) Drugs interacting with the herb: Anticoagulants drugs, antiplatelet drugs, immunosuppressants.[2a,b]

Contraindicated in bleeding disorders, hypotension, male infertility. Safety in pregnancy and lactation not established.[3]

Anethum graveolens DC.
Shataahvaa, Indian Dill

The extract of fruits may have teratogenic effects. The use of the fruit during pregnancy is not recommended.[16c]

Angelica archangelica Linn.
Chandaa, Wild Celery

Root contains furanocoumarins; intense UV radiation should be avoided.[1,2,10]

Emmenagogue, uterine stimulant,[10] can induce miscarriage,[7] can cause uterine contractions.[2a]

GRAS status in the U.S. Canada does not allow *Archangelica* species as food ingredients.[2b]

Angelica sinensis (Oliv.) Diels
Chinese Angelica

Contraindicated in bleeding disorders, lactation, pregnancy.[5]

Osthole and ferulic acid may inhibit platelet aggregration.[5]

Herb interacts with anticoagulants, antiplatelet agents, estrogen replacement therapy/oral contraceptives, may result in estrogen excess.[12]

Anisomeles malabarica (Linn.) R. Br. ex Sims
Sprukkaa, Malabar Catmint

Aqueous extract of the shoot: Spermicidal effect in albino rats *(in vitro)*, also on human semen.[4b]

Anogeissus latifolia Wall. ex Bedd.
Dhava, Axle-wood

Alcoholic extract of the stem bark: CNS depressant, hypothermic, responds to amphetamine hyperactivity test.[4d]

Apium graveolens Linn.
Ajmoda, Celery

Contraindicated in pregnancy and kidney disorders.

Furanocoumarin constituents may cause photosensitivity.[9]

Seed tablets interact with thyroxine, and anticogagulants. Potential allergenicity, including anaphylactic shock reported.[1]

Furanocoumarin content increases 100-fold in injured or diseased celery.[2a]

Aquilaria agallocha Roxb.
Aguru, Agarwood

Benzene extract of the wood exhibited potent CNS depressant activity in mice.[4c]

Areca catechu Linn.
Puuga, Betel Nut

Toxic constituents: Arecoline, arecain (pyridine alkaloids.)[11]

Interactions: High tannin of the nut may cause alkaloids to become insoluble and precipitate. The cholinergic activity of arecoline interacts with anti-cholinergic drugs.[5,14]

Asthmatic and chronic obstructive pulmonary disease (COPD) patients may be at risk.[14]

Argemone mexicana **Linn.**
Svarnkshiri, Mexican Poppy

Adulteration of edible oil with sanguinarine containing argemone oil has led to wide spread epidemic of dropsy and glaucoma. Sanguinarine is also carcinogenic.[4b]

Traditional medicinal uses, attributed to this plant are based on wrong identity.[4b]

Argyreia speciosa **Sweet.**
Vriddhadaaruka, Elephant-creeper

Seeds contain 0.5-0.9 per cent ergoline alkaloids; reported to be hallucinogenic.[4b]

Aristolochia indica **Linn.**
Ishvari, Indian Birthwort

Toxic constituents: Aristolochin, aristolochic acid.[11] Toxic to kidneys. Carcinogenic in animals and human cells.[2a, 11] Many cases of nephropathy associated with aristolochia use have been reported worldwide.[2a]

Any product which contains plants known or suspected to contain aristolochic acid, is detained in the U.S.[2a] Aristolochia is also banned in Germany, Austria, France, Great Britain, Belgium and Japan.[2b]

A cytotoxic lignan, savinin, has been isolated from the root.[4c] Root: Oxytocic, emmenagogue, abortifacient.[4b]

Artemesia nilagirica **(Clarke) Pamp.**
Damanaka, Wormwood

Emmenagogue, uterine stimulant.[4b,10] Toxic constituent: Thujone.[11,4c]

May interact with anticonvulsants, may lower seizure threshold.[12]

Artocarpus heterophyllus **Lam.**
Panasa, Jackfruit

A lectin, jacalin, a potent polyclonal activator for human lymphocytes, has been isolated from the seeds.[4c]

Artocarpus lakoocha **Roxb.**
Lakucha, Monkey Jack

A lectin, artocarpin, isolated from the seeds precipitates several galactomannans. Besides human and animal erythrocytes, it agglutinates rat lymphocytes and mouse ascites cells.[4c]

Asparagus officinalis **Linn.**
European Asparagus

Contraindicated in inflammatory kidney diseases; in edema due to functional heart or kidney disorders.[1,10]

Asparagus racemosus Willd.
Shaatavari, Indian Asparagus

Roots are reported to show inhibitory effects on digestive enzymes, lipase and trypsin; may lead to stoppage of degradation of food material in the intestinal tract.[4c]

Azadirachta indica A. Juss.
Nimba, Neem

The oil might cause morphological changes in sperm, leaf can effect sperm motility and viability. Avoid in couples with infertility.[2]

Might decrease the effectiveness of immunosuppresants in organ transplant patients.[2a]

The oil had embryotoxic effects after vaginal administration to pregnant rats at a dose of 0.25 ml/animal. Embryotoxic effects were also reported following intragastric administration of 4.0 ml/kg bw of the oil to pregnant rats on days 6-8 of pregnancy. Oral administration of the oil is contraindicated during pregnancy, nursing and in children under the age of 2 years. The oil has demonstrated antifertility effects in numerous animal and human studies.[16c]

A case of ventricular fibrillation and cardiac arrest due to Neem leaf poisoning has been reported. Owing to genotoxic effects, the leaves should not be administered during pregnancy or nursing, or to children under the age of 12 years.[16c]

Bacopa monnieri (Linn.) Penn.
Braahmi, Thyme-leaved Gratiola

Herb prolongs hypnotic effect of pentobarbitone; may antagonize the haloperidol-induced catalepsy, suggesting an involvement of GABA-ergic system. Potentiates phenothiazines.[5]

Herb's CNS action does not include serious sedation; caution with phenothinzine co-administration is indicated.[3]

Preparations of *Bacopa monnieri* are contraindicated with antidepressant or antipsychotic medications, thyroid medications, calcium blocking drugs. (https: // www.livestrong.com/article/58203-side-effects-bacopa-monnieri/.)

Bambusa arundinacea Willd.
Vansha, Spiny Bamboo

Tender shoots and root contain a cyanogenic glucoside.[4b.] Ethanolic extract of tender shoots adversely affects the sperm count and motility in rats.[4c]

Bauhinia variegata Linn.
Kaanchnaara, Mountain Ebony

Seeds exhibit haemagglutinating activity.[4c]

Berberis asiatica Roxb. ex DC.
Daaruharidra, Indian Barberry

Toxic constituent: Berberine (isoquinoline alkaloids.) May inhibit P450 enzymes.[5] (*Berberis aristata* stem bark yielded 2.76 per cent berberine, *Berberis vulgaris* up to 4.5 per cent.)

Drug interactions with *Berberis vulgaris*: Berberine is reported to upregulate the expression of the human multidrug resistance gene coding for multidrug resistance transporter (PGP-170); thus the treatment of tumours with berberine may result in reduced retention of chemotherapeutic agents such as paclitaxel.[16d]

Berberine has been reported to interact with cyclosporine in renal transplant patients. Blood concentrations of cyclosporine were enhanced by 75 per cent after co-administration of berberine hydrochloride in renal transplant patients, but this did not increase the toxicity of cyclosporine.[16d]

Bergenia ciliata Sternb.
Paashaanabheda (AFI)

Acetone extract of the rhizome is reported to be cardiotoxic in higher doses and depressant on CNS. Mildly diuretic; in higher doses anti-diuretic in experimental animals.[4b]

Boerhaavia diffusa Linn.
Punarnavaa, Hogweed

Official as a diuretic in IP. The cholinergic activity of the root was observed *in vitro* but not *in vivo*.[4d.]

Boswellia serrata Roxb. ex Colebr.
Shallaki, Indian Olibanum

Indian Frankincense might inhibit mediators of autoimmune disorders. It seems to reduce production of antibodies and cell-mediated immunity.[2b]

Brassica alba Boiss
Sarshapa Gaura, White Mustard

Isothiocyanates can cause endemic goitre.[2a] Irritant poisoning (on ingestion of a large quantity) can occur in people with kidney disorders.[2a] GRAS status in the US.

Brassica campestris Hook. f. & Thoms.
Sarshapa, Field Mustard

The pollen grains act as allergens causing bronchial and broncho-pulmonary problems. Glycoprotein containing 11.4 per cent of carbohydrate is the main allergen.

Brassica juncea (Linn.) Czern. & Coss.
Raajikaa, Brown Mustard

Toxic constituent: Glucosinolates.[11] Vegetable containing glucosinolates are goitrogenic.[5]

Drugs interacting with black mustard: Antacids, H_2 Blockers, Proton Pump Inhibitors.[2a]

Butea monosperma (Lam.) Taub.
Palaasha, Flame-of-the-forest

Flower and seed: Anti-estrogenic. Seed: Nephrotoxic. Anthelmintic principle: Palasonin.[4b]

Caloptropis gigantea (Linn.) R.Br. ex Ait.
Arka, Milk-weed

Root: Cardiac glycosides [2a] may be additive with digoxin.[5]

Drugs interacting with the herb: Digoxin, diuretic drugs, stimulant laxatives.[2a]

Canabis sativa Linn.
Vijayaa, Indian Hemp

Toxic constituents: Tetrahydrocannabinol (Cannabinoids 15-20 per cent)[11]

Oxytocic, crosses human placenta, high doses damage developing embryos.[11]

Carum carvi Linn.
Jeeraka Krishna, Caraway

Seed: May lower blood glucose and test results.[2a] Should not be used in gastro-esophageal reflux disease or during pregnancy (uterine relaxation may occur.) Side effects include renal dysfunction.[7]

Drugs interacting with the herbs: Antidiabetic drugs.[2a]

Subchronic and chronic toxicity: A study in rats demonstrated that 1 per cent of (+)-carvone in the diet for 16 weeks caused growth retardation and testicular atrophy, while 0.1 per cent for 28 weeks and 0.25 per cent for one year had no effects. WHO has established an ADI (acceptable daily intake) for (+)- carvone of 0.1 mg/kg bw/day.[8a] (WHO Geneva: Food Additives series 28, 1991: 155-167.)

Cassia senna Linn. var. Senna
Svarnpatri, Tinnevelly Senna

Contraindicated in abdominal pain of unknown origin, pregnancy, lactation, appendicitis, Crohn's disease, ulcerative colitis, ileus, intestinal inflammation or obstruction, children under 12 year of age;[1,5,8a,16a] in hemorrhoids.[10]

With chronic use or abuse: Electrolyte imbalance, potassium deficiency, albuminuria, hematuria.[1]

May potentiate toxicity of digitalis and diuretics.[12] Chronic herb use may increase effects of cardiac glycosides.[5]

Toxic constituents in senna spp.; anthraquinone glycosides, dianthrones, anthrones.[11]

Drugs interacting with the herbs: Digoxin, diuretic drugs.[2a] Herb may potentiate antiarrhythmics. Herb used with corticosteroids may cause hypokalemia.

Anthranoid metabolities of dried ripe fruits may lead to false positive test results for urinary urobilinogen and for estrogen measured by the Kober procedure.[16a]

Celastrus paniculatus Willd.
Jyotishmati, Staff Tree

Seed oil: CNS depressant, additive to pentobarbitol.[4b]

Centella asiatica (Linn.) Urban
Mandukaparni, Asiatic Pennywort

Emmenagogue.[3,6] Excessive internal use is contraindicated in early pregnancy.[6]

Canadian regulation do not allow the herb as a non-medical ingredient for oral use.[10]

Asiaticoside has been implicated as a possible skin carcinogen in rodents after repeated topical application. Further experimentation is needed to substantiate this claim.[16a]

Chenopodium album Linn.
Vaastuka, Lamb's Quarter

Toxic constituents of *C. ambrosioides* (American wormseed): Ascaridole (unsaturated terpene peroxide of volatile oxide.) Emmenagogue, abortifacient.[11]

Cinnamomum camphora (Linn.) Nees and Eberm.
Karpura, Camphor Tree

Contraindicated externally on injured skin, burns,[1] preparation not to be used in the facial region of infants and small children, especially in the nasal area.[1,10] Not for long term use.[10]

Toxic constituents: camphor (volatile saturated ketone), 30-50 per cent of volatile oil. Safrole, cincole, (volatile oxides) in crude camphor oil.[11]

Camphorated oil (20 per cent camphor in oil) was removed from the US market due to toxicity in 1980s. Available without a prescription in Canada.[2a]

Cinnamomum zeylanicum Blume
Tvak, Ceylon Cinnamon

Contraindicated in pregnancy,[1] stomach or duodenal ulcers.[16a]

There is one report of teratogenicity of cinnamaldehyde (0.5 mmol/embryos) in chick embryos (58.2 per cent malformations and 49 per cent lethality), while a menthol extract of the drug given by gastric intubation was not teratogenic in rats.[8a,16a]

Large doses have caused methemoglobinemis, hematinemia with nephritis and stimulation of vasomotor center.[5,10]

Drugs interacting with the herb: Antidiabetic drugs.[2a]

Cissampelos pareira Linn.
Paathaa, Pareira Brava

Root: Alkaloid haytine (methiodide and methochloride derivatives), reported to be potent neuromuscular-blocking agent.[4b]

Cissus quadrangularis Linn.
Asthisamhaara, Square-stalked Vine

Total alcoholic extract of the plant, on parenteral administration, neutralizes antianabolic effect of the cortisone in healing fractures.[4b]

Citrullus colocynthis Schrad.
Indravaaruni, Colocynth Bitter Apple

Contraindicated in infections or inflammatory gastrointestinal conditions.[2]

Drugs interacting with the herb: Digoxin, diuretic drugs.[2]

Clerodendrum phlomidis Linn.
Agnimantha, Arni Tree

Plant exhibited antiovulatory activity and prevented pregnancy in albino rats.[4c]

Coix lacrymal jobi Linn.
Gavedhukaa, Job's Tears

Seeds and leaves: May promote fertility in woman; trans-ferulyl stigmastanol and trans-ferulyl campestanol (isolated from the seed) may induce ovulation.[4c]

Commiphora mukul Hook. ex Stocks
Guggulu, Indian Bdellium

Emmenagogue, uterine stimulant.[10]

In contrast to studies of guggulu in Indian population, taking guggulu in dose of 3000 or 6000 mg per day does not seem to lower total cholesterol or triglycerides, or raise HDL cholesterol in people on western diet. It seems to increase LDL cholesterol by 9 to 10 per cent [2] (*JAMA*, 2003, 290: 765-72.)

Interactions: Gugulipid decreased peak serum concentration and area under curve of propranolol. Gugulipid decreased peak serum concentration and area under the curve of diltiazem. Z-guggulsterone may increase uptake of iodine by thyroid gland and increase oxygen uptake in liver and bicep tissues.[5]

Commiphora myrrha (Nees.) Engl.
Bola, Myrrh

Amounts greater than 2-4 g can cause kidney irritation and diarrhoea; Large amounts can affect heart rate.[2a] Internal ingestion of the gum may interfere with existing antidiabetic therapy owing to the ability of the resin to reduce blood glucose levels.[16c]

Crocus sativus Linn.
Kumkuma, Saffron

Toxic constituents in stigma and styles: Alpha-crocin (carotenoids), picrocrocin (volalile glycoside.)[11] Emmenagogue, uterine stimulant, abortifacient,[10] Side effects occur on 5 g, lethal dose 20 g, therapeutic dose 1.5 g/day.[10] Abortifacient dose 10 g.[1]

Side effects include severe purpura, a thrombocytopenia of 24,000, hypothrombinemia of 41 per cent, severe collapse with uremia, bleeding from uterus, bloody diarrhea, bleeding from nose and eyelids, vertigo, dizziness, numbness.[1,16c]

Croton tiglium Linn.
Jayapaal, Purging Croton

Toxic constituents in oil from seeds: Phorbols (terpenoids) from nonvolatile oil. Crotin, a toxic albuminous substance, is not extracted in the oil.)[11] Phorbol esters: Tumour promoting.[4c]

Curcuma longa Linn.
Haridraa, Turmeric

Root: Contraindicated in obstruction of bile passage, in cases of gallstones use only after consulting a physician.[1,5,10,16a] Not to be administered to patients of stomach ulcers or hyperacidity.[10,5] Emmenagogue, uterine stimulant.[10,5]

Curcumin may potentiate antiplatelet activity, interacts with antiplatelet agents.[12]

Synonyms Rajani, Nishaa, Nishi, Raatri, Nilkanth should be equated with *Curcuma caesia* Roxb. (C. P. Khare.)

Cymbopogon citratus (DC.) Stapf.
Bhuutikaa, Lemongrass

Contraindicated in glaucoma, (citral raised ocular pressure in very low doses in experimental studies), in benign prostatic hyperplasia (due to citral),[6] in pregnancy.[5] Emmenagogue, uterine stimulant.[10]

Adverse reactions include slight elevation of direct bilirubin and amylase.[2,5]

Datura metel Linn.
Dattuura, Thorn Apple

Toxic constituents: Hyoscine, hyoscyamine.[11] Adverse reactions: Agitation, blurred vision, dilated pupils, disorientation, hallucinations, mydriasis, tachycardia, seizures, xerostomia and coma.[5] Lethal dose for adults: 15-100 g dry leaf powder; 15-25 g seeds; 100 mg atropine.

Desmodium gangeticum DC.
Shaaliparni

Root extract exhibited significant antifertility activity at 50 and 100 mg/kg in rats.[4c]

Elettaria cardamomum Maton.
Elaa, Lesser Cardamom

Contraindicated in case of gallstones (to be used only after consultation with physician),[1] in gastroesophageal reflux disease.[7]

Drug interaction: Application of the essential oil from the seed at a concentration of 1.0 per cent to rabbit skin enhanced the dermal penetration of piroxicam, indomethacin and diclofenac sodium.[16d]

Embelia ribes Burm. f.
Vidanga, Embelia

Embelin, isolated from berries: Antispermatogenic. Changes were found to be reversible.[4c]

Ephedra gerardiana Wall. ex Staph.
Soma (one of the constituents), Ephedra

Toxic constituents: Ephedrine, pseudoephedrine, (2-aminophenylpropane alkaloids.)[11]

Contraindicated in anxiety, high blood pressure, glaucoma, impaired circulation of the cerebrum, adenoma of the prostate with residual urine accumulation, pheochromocytoma, thyrotoxicosis.

May potentiate MAO inhibition.[3, 10, 13] Interacts with methyldopa, beta-blockers, caffeine, monoamine oxidase inhibitors, theophylline sympathomimetics, St John's Wort, Guanethidine, cardiac glycosides, oxytocin.[12]

Euphorbia nerrifolia auct. non Linn.
Snuhi, Dog's Tongue

Canadian regulations do not allow *Euphorbia* in foods.[10]

Ferula foetida Regel.
Hingu, Asafoetida

Emenagogue, uterine stimulant.

Contraindicated in infant colic,[10] in bleeding disorders, pregnancy, inflammatory GI diseases,[5] peptic ulcers.[6] 50-100 mg of gum resin may cause convulsions in persons suffering from nervine diseases.[2a,10]

Drugs interacting with the herb: Anticoagulant/antiplatelet drugs, antihypertensive drugs.[2a]

Ficus benghalensis Linn.
Nyagrodha, Banyan Tree

Bark extracts, containing leucodelphinidin and bengalenoside, not to be taken concurrently with diabetes medications.[5] These constituents decreased blood glucose in research animal.[5]

Foeniculum vulgare Mill.
Mishreya, Fennel

Owing to the potential estrogenic effects of the essential oil from seeds and anethole, its traditional use as an emmenagogue is contraindicated. Seeds due to lack of human studies are also contraindicated in pregnancy.[16c] Fennel oil not to be used for infants and toddlers[1.] Estragole (5-10 per cent of essential oil) is a procarcinogen.[10] Concomitant use of fennel may reduce ciprofloxacin bioavailability by nearly 50 per cent possibly due to calcium, iron and magnesium contained in fennel. Fennel increases tissue distribution and slows elimination of ciprofloxacin.[2a, 14]. Similar effects may be expected from fluroquinolones.[14.]

Fumaria parviflora Lam.
Parpata, Fumitory

Should not be used by persons with seizure disorders or increased intraocular pressure.[7]

Fumaria officinalis contraindicated in biliary obstructions. In cases of gall stones, not to be used without medical advice.[8b.]

Gentiana spp. Linn.
Gentian Root

Toxic constituents of *G. lutea*, yellow Gentian. root: Amarogentine, gentiopicrin, swertiamarin, sioeroside, (iridoid monoterpenes.)[11] Contraindicated in gastric and duodenal ulcers.[1]

Glycyrrhiza glabra Linn.
Yashtimadhu, Licorice

Toxic constituent in the root: 6-14 per cent Glycyrrhizic acid (saponin glycoside.) Interacts with spironolactone (antagonises diuretic effect), cardiac glycosides, thiazide diuretics (can cause hypokalemia), digoxin (may potentiate toxicity), corticosteroids, cyclosporine, monoamine oxidase inhibitors (immunostimulating effect may decrease response to the drugs.) Sympathomimetic amines may increase the risk of hypertensive crisis.[12,10]

Contraindicated in cholestatic liver disorders, liver cirrhosis, hypertonia, hypokalemia, severe kidney insufficiency, pregnancy,[1] bleeding disorders, diabetes (on insulin), impotence, male infertility.[5] Deglycyrrhizinized licorice (DGL) is usually free of side effects.[10] French regulation limits daily consumption to 5 g or as tea 8 g.[10]

Prolonged use (>6 weeks) of excessive doses of the root (>50 g/day) can lead to pseudoaldosteronism, which includes potassium depletion, sodium retention, edema, hypertension and weight gain.[16a]

The root should not be taken concurrently with corticosteroid treatment.[16a]

Gossypium herbaceum Linn.
Kaarpaasa, Asiatic Cotton

Root bark: Abortifacient, emmenagogue, uterine stimulant. Contraindicated in urogenital irritation or tendency to inflammation. Chronic use may cause sterility in man.[10]. Gossypol decreases sperm count, interacts with diuretics and potentiates hypokalemia, increases gastrointestinal irritations. Concurrent use with thyroid replacement therapy may require an increase in dosage of the drug.[5]

Canadian regulation do not allow an excess of 450 ppm of free gossypol in foods (cotton seed meal and oils.)[10.] Drugs interacting with gossypol: Digoxin, diuretic drugs, NSAIDS, stimulant laxatives, theophylline.[2a]

Hemidesmus indicus R. Br.
Anantamuula, Indian Sarsaparilla

Known as Indian Sarsaparilla. According to Tyler (*Honest herbals*), does not contain same saponins or other principal constituents found in Sarsaparilla. Sarsaparilla of Western herbals may increase absorption of Bismuth drug; Digoxin and cardiac glucosides. It may increase excretion of simultaneously taken hipnotics.[5]

Hordeum vulgare Linn.
Yava, Barley

Seeds contain gliadin, a component of gluten.[5] May supress the secretion of milk in women whose child have died after birth. (*Chinese Materia medica.*)

Hyoscyamus niger Linn.
Paarsika-yavaani, Black Henbane

Toxic constituents in whole plant: Scopolamine (hyoscine), hyoscyamine, butyrine (tropane alkaloids), hyospicrin (glucoside.)[11] Leaf contraindicated in tachycardiac arrythmias, prostatic adenoma with urine retention, narrow angle glaucoma, acute pulmonary edema, mechanical stenosis in any part of the gastrointestinal tract, megacolon.[1] Hyoscymine and scopolamine alkaloids are anti-cholinergic.[5]

Imperata cylindrica Beauv.
Darbha, Thatch Grass

Leaves and stem contains cyanophoric compounds.[4c] Plant can be used as a substitute for Ephedra (as an anti-fatigue agent.)[4c]

Inula racemosa Hook.
Pushkarmuula, Elecampane

The root powder has been found to possess beta-blocking activity.[4c]

Juniperous communis Linn.
Hapushaa, Common Juniper

Fruits are contraindicated in pregnancy and inflammatory kidney diseases.[1,10] Not for use exceeding 4 to 6 weeks in succession. Canadian regulation do not allow juniper as a non-medical ingredient for oral use products.[10] The volatile monoterpenes are irritant to the urinary mucosa.[11]

Lathyrus sativus Linn.
Khesaari, Chickling Vetch

Seeds: Neurotoxic.[4c]

Lawsonia inermis Linn.
Madayanti, Henna

Leaf: History of internal use as an abortifacient is recorded in Africa.[10] Leaf juice (50 g) is taken as an oral herbal contraceptive in both the sexes. Prolonged use may cause permanent sterility.[4c.]

Linum usitatissimum Linn.
Atasi, Flax

Contraindicated in intestinal obstruction of any origin,[1,8a] bleeding disorders, elevated prolactin, hypokalemia.[5] Use during pregnancy or lactation, only on medical advice.[5] Toxic constituents linatine, (glutamic acid derivatives) from seed; lotaustralin (cyanogenic glycosides) from leaves, stems and roots; 0.1-1.5 per cent linustatin and neolinustatin (cyanogenic glycosides) from seeds.[11] In spite of cyanogenic glycosides, single dose of up to 150-300 g of powdered linseed are not toxic.[8a]

Defatted flaxseed meal may decrease zinc status, as shown by decreased alkaline phosphatase. Defatted flaxseed may increase serum triglycerides.

Mucilage content may decrease absorption of other herbs. Flaxseed oil may decrease platelet aggregation, which may interact with anticoagulants and increase the risk of bleeding. Flaxseed has been shown to alter metabolism of endogenous hormones as well as increase serum prolactin in postmenopausal women. Effect on exogenous hormones and clinical significance is not known. It interacts with hormone replacement therapy and oral hypoglycemic agents.[5]

Madhuca indica J. F. Gmel.
Madhuuka, Mahua Tree

Seed oil causes total but reversible sterility in male rats; shows testicular atrophy and degeneration of seminiferous tubules.[4c]

Maranta arundinacea Linn.
Tugaa, Arrowroot

The tuber contains a protease inhibitor, which inhibits the proteolytic activity of human and bovine pancreatic secretion.[4d]

Melia azedarach Linn.
Mahaanimba, Pride of India

Cases of severe poisoning due to fruits have been reported.[4a]

Mimosa pudica Linn.
Lajjaalu, Sensitive Plant

Feeding trials with horses in large doses caused toxic symptoms including loss of hair. Leaf meal caused stunted growth in Chicks.[4a]

Momordica charanita Linn.
Kaarvallaka, Bitter Gourd

Potentiates effects of hypoglycemic drugs.[12] Safety in pregnancy is not established. Red arils should not be consumed by children. Excessive amount can cause diarrhoea and GI distress.[5] Juice: Emmenagogue and abortifacient.[6]

Moringa oleifera Lam.
Shigru, Drumstick Tree

Root and bark extract: Antifertility activity in experimental animals.[4c] Root more potent than bark.[4c] The leaf extract may have anti-thyroid effects. A study in animal models reported reduced serum triiodothyronine (T_3) levels and increased thyroxine (T_4) levels, suggesting reduced peripheral conversion of T_4 to T_3.[2b]

Myrica nagi Hook. f. non Thunb.
Katphala, Box Myrtle

Canadian regulations do not allow bayberry (Myrica) as a non-medical ingredient for oral use products.[10] Fruits are sedative.[4a]

Myristica fragrans Houtt.
Jaatiphala, Nutmeg

Toxic constituents: Myristicin, elemicine, safrole. More than 5 g of powdered nutmeg, or mace affects the central nervous system, producing hallucinations, headache, dizziness, drowsiness, nausea, stomach pain, liver pain, excessive thirst, rapid pulse, anxiety, double vision and sometime acute panic, coma or death.[1,5,10] With ingestion of 9 teaspoons of nutmeg powder per day, an atropine like effect was observed.[1,5] When taken in larger amount, the herb has abortificient action.[1]

Nardostachys jatamansi DC.
Jatamansi, Spikenard

Rhizome: Emmenagogue, uterine stimulant.[10] Oil: Potentiated phenobarbital narcosis, reduced brain serotonin content in rats.[4c] Action at variance with that of reserpine which has direct action on cell to liberate serotonin.[4c]

Nerium indicum Mill.
Hayamaaraka, White Oleander

Powdered extract of leaf: Contraindicated in hypercalcemia.[5] Entire plant is toxic, caused fatal poisonings. Contains cardiac glycosides.[5] Between 65-86 per cent of the cardioactive glycosides are absorbed and about 50 per cent are protein bound. The duration of effect is about 2.65 days.[11]

Ocimum sanctum Linn.
Tulasi, Holy Basil

Essential oil contains estragole (a procarcinogen.)[5]

Interactions with drugs: Anticoagulants and drugs that increase the risk of bleeding. Fixed oil of herb increased clotting time and possibly may inhibit platelet aggregation, as seen in laboratory animals.

Bromocriptine: In mice, ethanolic extract of herb, when combined with Bromocriptine, had a synergistic effect, which suggests that herb may have D2 agonist activity.

Doxorubicin (Adriamycin): In *in vitro* research, ursolic acid from herb showed a 13 per cent protection of liver cells and 17 per cent protection of heart cells.

Thyroid medications: In mice, leaf extract decreased serum T_4. Herb leaf extract to be used with caution in thyroid diseases.[5]

Papaver somniferum Linn.
Ahiphena, Opium Poppy

Toxic constituents in latex from unripe seed capsules: Morphine, codeine, papaverine (isoquinoline alkaloids.)[11] Opium is a controlled substance in many countries.[10]

Picrorhiza kurroa Royle ex Benth.
Katukaa, Picrorhiza

There is some concern that *Picrorrhiza* might adversely affect patients with autoimmune disorders because of its immune stimulating effects. Avoid using in patients with multiple sclerosis, systemic lupus erythematosus, rheumatoid arthritis and other autoimmune disorders.[2] Interacts with Immunosuppresants.[2]

Piper cubeba Linn.
Kankola, Cubeb

Contraindicated in nephritis,[2a,10] also in individuals with infections or inflammatory GI conditions.[2] Drugs interacting with the herb: Antacids, H_2 Blockers, Proton Pump Inhibitors.[2a]

Piper longum Linn.
Pippali, Indian Long Pepper

Contraindicated in pregnancy in large doses.[5,6] Piperine inhibits gastric emptying and GI transit. Also increases drug-induced sleep.[5]

Piper nigrum Linn.
Maricha, Black Pepper

Contraindicated in pregnancy in large doses.[5,6].

Phenytoin was more rapidly and more completely absorbed and eliminated more slowly when taken with piperine.[6]

Piperine can augment the therapeutic effects of phenobarbital, phenytoin, rifampicin, theophylline and other drugs, it can also predispose the patient to an increased risk of side effects of drugs with a narrow therapeutic window that are liver metabolized (*e.g.*theophylline.)[14.]

Drugs interacting with the herbs: CYP 3A4 substrates, P-glycoprotein substrates, Phenytoin, Propranolol, Rifampicin, Sparteine, Theophylline.[2a,5]

Pisum sativum Linn.
Matar, Pea

An antifertility agent, m-xylohydroquinone-one, isolated from peas, produced fatal resorption in rats; Trials on women with capsules containing 300-350mg of m-xylohydroquinone-one, twice a month, for variable periods showed 50-60 per cent reduction in pregnancy rate. In trial on men, the drug caused 50 per cent reduction in the number of spermatozoa.[4a,d]

Polygonatum verticillatum Linn.
Medaa (of Ashtavarga, recently identified), Himalayan Solomon's Seal

Plant contains a cardiac glucoside.[4a] Rhizomes may be used as a source of diosgenin.[4c]

Prunus amygdalus Batsch
Vaataama (Sweet var.), Almond

A Japanese patent claims isolation of low molecular weight peptides with analgesic and anti-inflammatory properties.[2a] Bitter almonds (var. *amara*) is toxic orally due to the benzaldehyde and hydrocyanic acid content. Ingesting the bitter kernel or kernel oil can cause fatal CNS depression with respiratory failure.[2b]

Psoralea corylifolia Linn.
Baakuchi, Purple Fleabane

Toxic symptoms of seeds: Nausea, vomiting, malaise, GI disturbances. External application of oil: Irritant to skin, causes blistering.[4a] Oil even in high dilutions (up to 1 in 100,000) increases tone of the uterus and stimulates smooth muscles of experimental animals.[4a] A mixture of psoralen and isopsoralen (in 1:3 ratio) is recommended in leucoderma.[4d]

A powder has been patented in China containing salt as the main ingredient with *P. corylifolia* extract for odontopathy, dental caries and periodontal diseases. In Japan, oral compositions are being patented containing extracts of *P. corylifolia* for controlling *Streptococcus mutans* related dental caries.[4d]

Punica granatum Linn.
Dadima, Pomegranate

Dried root bark, stem bark, seeds contain: toxic constituents pelletrin, piperidine alkaloids 0.4 per cent, punicalagin, punicacontein C, casurin, and tannins 20-25 per cent.[10, 11]

Quercus infectoria Oliv.
Maajuphala, Oak Gall

Oak bark: Contraindicated in cardiac insufficiency, hypertonia and externally on broken skin.[5] Tannins may interfere with absorption of drugs.[5]

Galls contain 50-70 per cent gallotannic acid, 2-4 per cent gallic acid, nyctanthic acid, rubric acid. (Diluted powder of Galls is used in leucorrhoea, dry and itching vagina, vaginal laxity and topically for dental inflammation. API.)

Randia dumetorum Lam.
Madanaphala, Common Emetic Nut

Fruit pulp: Emetic and abortifacient; potentiates pentobarbitone, depresses heart, relaxes ileum, antagonises effect of acetylcholine (in animal studies.)[4a,c]

Raphanus sativus Linn.
Muulaka, Radish

Whole plant is contraindicated in cholelithiasis.[1] It might cause biliary colic.[2a]

Rauvolfia serpentina (L.) Benth. ex Kurz
Sarpagandhaa, Indian Snakeroot

Contraindicated in depression, ulcer, pheochromocytoma, pregnancy and lactation,[1] also in Parkinson's disease.[7] Toxic constituents: Reserpine, serpentine, raupine, ajmaline and other indole alkaloids.[11]

Interactions with drugs: Digitalis glycosides or quinidine, levodopa, levomepromazine, monoamine oxidase inhibitors, sympathomimetics (direct-acting), tricyclic antidepressants, alcohol or other central nervous system depressants, other antihypertensives and diuretics.[1,16a]

Reserpine administered parenterally has been shown to be teratogenic in rats at doses up to 2 mg/kg and to have an embryocidal effect in guinea-pigs at 0.5 mg daily. There are no well-controlled studies in pregnant women.[16a]

Rheum emodi Wall.
Amlaparni, Indian Rhubarb

Rhubarb root: Contraindicated in intestinal obstruction, Crohn's disease, colitis ulcerosa, appendicitis, abdominal pain of unknown origin, pregnancy and children under 12 years of age.[1,16a]

Ricinus communis Linn.
Eranda, Castor Seed

Seed oil: Contraindicated in intestinal obstruction and abdominal pain of unknown origin, appendicitis, ulcerative colitis.[5,10.] Must be used with caution during pregnancy and menstruation.[10] Toxic constituents: 3 per cent ricin (toxalbumin), ricinine.[11] (Ricin as a warfare agent: weapons grade ricin is purified and produced in inhalable particles that can be aerosolized for a mass attack.[2a])

Salvia plebeia R. Br.
Samudrashosha, Sage Weed

Sage leaf: Essential oil and alcoholic extracts contraindicated during pregnancy.[1,10] Sage oil contains more thujone than absinthium oil, yet it has not been reported as toxic.[10] Alcoholic preparations to be used with caution due to the presence of thujone.[8a]

Santalum album Linn.
Shveta Chandana, White Sandalwood

Oil is considered to be the kidney irritant.[2a] Use longer than 6 weeks not advised.[2a] Extracts of the alpha- and beta-santols are sedative and could be considered as neuroleptic by resemblance to pharmacological activities of chloropromazine.[4d]

Saussurea lappa C.B. Clarke
Kustha, Kuth

The root is commonly contaminated with aristolochic acid which is nephrotoxic and carcinogenic. Any product which contains plants, known or suspected to contain aristolochic acid is detained in the US.[2b]

Sida cordifolia Linn.
Balaa, Country Mallow

Ephedrine is reported to be present in the plant [4a.]

Drugs interacting with the herb: Antidiabetic drugs, dexamethasone, ergot derivatives, methylxnathine, MAOs, tricyclic antidepressants.[2b]

Spinacia oleracea Linn.
Paalankya, Spinach

Spinach is not recommended in diets of growing children, pregnant women and patients suffering from calcium deficiency (due to high content of oxalate in leaves.)[4a] (Oxalates can be eliminated by boiling the vegetable for 15 minutes and rejecting the water. Discarding the cooking water reduce nitrate content.[4a]

Strebulus asper Lour.
Shaakhotaka, Siamee Rough Brush

The Central Drug Research Institute, Lucknow, India, has developed an antifilarial elephantiasis drug (glycoside asperoside- K029 and glycoside streblloside-K030) from the crude extracts of the stem.[4c]

Strychnos nux-vomica Linn.
Vishamushti, Nux-vomica

Toxic constituents of dried ripened seeds: 1-2 per cent, strychnine and brucine, vomicine and other indole alkaloids, glycoside loganin, caffeotannic acid.[11] Toxic dose: 5mg strychnine.[11]

Swertia chirata (Roxb. ex Flem.) Karst.
Kiraatatikta, Chiretta

May exacerbate duodenal ulcers.[2a] Unlike most other bitters, it does not constipate the bowels, may nauseate in over doses.[4a] Hypoglycemic due to xanthone derivatives.[4c]

GRAS status in the U.S.

Syzygium aromaticum (Linn.) Merrill and Perry
Lavanga, Clove

Contraindicated in bleeding disorders.[5]

Smoking cloves may cause haemoptysis and irritation of mucous membrane.[5] (Clove cigarettes generally contain 60-80 per cent tobacco and 20-40 per cent ground clove.)[2a]

Eugenol and acetyl eugenol in clove oil inhibit platelet aggregation, which may be additive with anticoagulants.[5] Acute liver toxicity, intravascular coagulation, acidosis, CNS depression and coma have been reported.[5] Eugenol and caryophyllene had a narcotic effect after intravenous administration of high doses (200-400 mg/kg bw), and a sedative effect after intragastric administration of low doses (1-100 mg/kg bw) to mice.[16b]

Drugs interacting with the herbs: Anticoagulant/antiplatelet drugs.[2]

Syzygium cumini Linn. Skeels
Jambu, Java Plum

Seeds extract exhibited protection against the effect of exposure to gamma radiation (*J Radiat Res,* Tokyo, 2005, 46: 59-65.)[2a] Fruit extract not to be used concurrently with Aspirin or Ibuprofen.[2a]

Taxus baccata Linn.
Thuner, European Yew

The cancer chemotherapeutic compound taxol (paclitaxel) from the bark of *T. brevifolia* and semisynthetic docetaxel from *T. baccata* are not present in nature in effective therapeutic quantities. Taxol in taxus media cultivars "Hicksy" fresh needles is at 0.0086-0.0094 g per cent and in microwave and oven dried needles at 0.0052 and 0.0029 g per cent respectively.[11] (Himalayan Yew is equated with *Taxus wallichiana* Zucc., syn. *T. baccata* Linn. subsp. *wallichiana* (Zucc.) Pilgoe, *T. baccata* Hook. f.)

Terminalia arjuna (Roxb.) **Wight & Arn.**
Arjuna

Crude extract of the bark showed anti-implantation as well as foetus absorption activity; ethanolic extract showed only anti-implantation activity. The drug had no effect on spermatogenesis.[4d]

Terminalia chebula Retz.
Haritaki, Chebulic Myrobalan

Fruit: Contraindicated in acute cough, acute diarrhoea, early stage dysentery.[10]

Dietary administration of the fruit to rats, as 25 per cent of the diet, produced hepatic lesions which included centrilobular vein abnormalities and centrilobular sinusoidal congestion. Marked renal lesions were also observed, and included marked tubular degeneration, tubular casts and intertubular congestion. (The median lethal dose of a 50 per cent ethanol extract of the fruit was 175.0 mg/kg bw after intraperitoneal administration.)[16d]

Thevetia neriifolia Juss. ex **Steud.**
Karavira (Pita), Yellow Oleander

The oral absorbability of peruvoside is comparable to that of digoxin. It has very low toxicity. Peruvoside containing drug, Endocordin, has been marketed in Germany for cardiac insufficiency.[4d]

Tinospora cordifolia (Willd.) **Miers ex Hook. f. Thoms.**
Guduchi, Heart-leaved Moonseed

Interacts with insulin and oral hypoglycemic agents.[5] Experiments conducted on rabbits indicate that aqueous and alcoholic extracts caused reduction in fasting blood sugar, and glucose tolerance was increased, but a deterioration in tolerance occurred after a month's treatment.[4a] Maximum hypoglycemic effect of aqueous extract was found at 400 mg/kg/d and at third week of administration.[5]

Trachyspermum ammi (Linn.) **Sprague.**
Yavaani, Ajovan caraway

Thymol, produced from ajowan oil, is toxic in high doses, may lead to fatal poisoning.[4a]

Trapa natans Linn.
Shringaata, Water Chestnut

Dentrifrices, containing collagenase inhibitor extracted from the seed, are used for periodontal disease control.[4d] The collagenase inhibitor is anti-inflammatory, can be incorporated into the skin lotions and creams.[4d]

Tribulus terrestris Linn.
Gokshuraka, Puncture Vine

Hormonal activity is attributed to protodioscin constituent of *T. terrestris*.[2a] (*J Ethnopharmacol*, 2005, 96: 127-32.) In Bulgeria, plant is used for impotency. A pharmaceutical preparation developed from the plant, *Tribestan*, has been found to increase the libido, the number and motility of sperm in men; in women it improves ovarial functions.[4d] (Steroidal saponin protodioscin, 0.245 per cent, may be present in fruits only in specific geographical regions.)

Plant can cause neurotoxicity when used in high doses over a long period.[4d]

Trigonella foenum-graecum Linn.
Methi, Fenugreek

Contraindicated with all medications, may reduce absorption if used concurrently. Not to be used concurrently with anti-coagulants.[6]

The crude steroidal extract of seeds exerts both antifertility and antiandrogenic activity in male albino rats; fed orally 100 mg/day for 60 days significantly declined the sperm count.[4c] Owing to its effect on blood glucose levels in diabetic patients, seeds should only be used in conjunction with oral antihyperglycaemic agents or insulin under the supervision of a healthcare professional.[16c] Owing to its stimulatory effects on the uterus, the seeds should not be used during pregnancy.[16c]

Urginia indica (Roxb.) Kunth.
Vajrakanda, Indian Squill

White squill: Contraindicated with digitalis glycosides, potassium deficiency,[1] in hypercalcaemia and hyperkaelemia.[5] Toxic constituents are scillarin A and B, proscillaridin A (bufadienolides) and other glycosides.[11] New York Heart Association in stage I and II heart disease: 100-500 mg standardized squill bulb powder per day.[2a] Extended use of *Urginea maritama* increases effectiveness and side effects of glucocorticoids.[5]

Valeriana wallichi DC.
Tagara, Indian Valerian

The co-occurrence of three cyclopentane-sesquiterpenoids (valerenic acid, acetoxyvalerenic acid, and valerenal) is confined to *V. officinalis* and permits its distinction from *V. edulis* and *V. wallichii*.[16a]

Valerian: Interacts with opiates, alcohol, CNS depressants.[12, 5]

Vetiveria zizanioides (Linn.) Nash.
Ushira, Vetiver

Root: Emmenagogue, uterine stimulant. Regulated in the US as an allowable flavouring agent in alcoholic beverages only.[10]

Vitex agnus-castus Linn.
Nirgundi, Monk's Pepper Tree

Vitex agnus-castus berry: Contraindicated in depression associated with reduced estrogen level.[6] Not to be used during pregnancy and lactation.[7] Mutual attenuation effects might occur in patients under concomitant treatment with dopamine receptor antagonists.[8a,12] *Vitex agnus-castus* has dopaminergic effect, may antagonise effects of dopamine receptor antagonists (phenothiazines.)[12]

Possible harmful interaction: Bromocriptine and related drugs; may augment prolactin inhibitory effects.[14] Drugs interacting with *V. agnus-castus*: Antipsychotic drugs, contraceptive drugs, dopamine agonists, estrogens, metoclopramide,[2a] hormones that affect the pituitary.[5,16d]

Adverse reactions have been reported in some clinical trials. A review of 30 studies involving 11506 subjects reported total 246 events (approximately 2 per cent.) The major reactions reported included acne, changes in menstrual cycle, dizziness, gastrointestinal distress, increased menstrual flow, nausea, skin reactions, urticaria and weight gain. Minor adverse events include fatigue, hair loss, increased intraocular pressure, palpitations, polyuria, sweating and vaginitis.[16d]

Vitis vinifera Linn.
Go-stani, Wine Grape

Red vine leaf, aqueous dry extract (4-7: 1): Used for chronic venous insufficiency.[8b] Grape seed extracts are standardised to 85-95 per cent procynacidins.[7]

Herb-drug interactions: Procyanidolic oligomer (extracted from grape seeds) may inhibit the metabolism of drugs which are substrates of CYP1A2 and CYP3A4; flavonoids from purple grape juice decreased platelet aggregation and increased platelet-derived nitric oxide release and superoxide production, increased risk of bleeding with high doses.[5] Grape polyphenols extracted from skin and seeds decreased hepatic injury from alcohol (in research using rats), but no effect on ethanol-induced lipid changes was observed.[5]

Withania somnifera (Linn.) Dunal.
Ashwagandha, "Indian Ginseng"

Preliminary evidence: Ashwagandha might boost thyroid hormone synthesis and orsecretion. (*American Herbal Pharmacopoiea*, 2000: 1-25.) May increase serum T_4.[2,5] Contraindicated in pregnancy.[5,6,10.] May potentiate barbiturates.[10]

Possible harmful interaction with drugs: antipsychotic agents, benzodiazepines, carbamazepines, phenobarbital, phenytoin, primidone, valproic acid, tricyclic antidepressants, zolpidem.[13]

Ashwagandha of the classical Ayurvedic period was a different drug (that species is now extinct.) It was a reputed drug for impotency and tuberculosis (API.)[4a]

Zingiber officinale Rosc.
Aardraka (fresh), Shunthi (dry.) Ginger

Contraindicated in gall stones, (only to be used after consultation with a physician)[1] Safety of large doses in pregnancy not established[5] due to emmenagogue and abortifacient effects.[6] For nausea and vomiting in pregnancy: to be used only under medical supervision.[8b] Teratogenicity studies reported that rat foetuses exposed to high-dose ginger tea were heavier and had more advanced skeletal development than controls. Embryonic loss was greater in the treatment group. In another similar study with different type of rats, no teratogenicity was observed.[3] FDA considers ginger as GRAS.[3]

Drugs interacting with the herb: Calcium channel blockers.[2a]

References

1. *The Complete German Commission, E Monographs*, Blumenthal *et al.*, American Botanical Council, Austin, 1998.

2. *Natural Medicines Comprehensive Database*, Therapeutic Research Faculty, Stockton CA, 2007(a); 2013 (b.)

3. *The Review of Natural Products*, Wolters Kluwer, 7th Edition, 2012.

4. *The Wealth of India*, CSIR, Original Series, Volume II to XI: (a); Revised Vol 1 to 3: (b); First supplement Series, Vol-I to 5: (c); Second supplement series Vol-1 to 3: (d.)

5. *Herb Drug Interaction Handbook*, Sharon M Herr, Church Street Books, Nassau, NY, Second Edition, 2002.

6. *Herb Contraindications and Drug Interactions*, Francis and Brinker, Eclectic Medical Publications, Sandy, Oregon, 2001.

7. *Handbook of Herbs and Naturals Supplements*, Linda Skid More, Roth Mosby, St Lois, Missourie, 2004.

8. *ESCOP (European Scientific Co-operative on Phytotherapy), Monographs*, Thieme, Stuttgart, Germany, Second Edn, 2003, 8a; Second Edn Supplement, 2009, 8b.

9. *Herbal Medicinals: A Clinician's Guide*, Lucinda G Miller and Wallace Murray, Haworth Press, New York, 1998.

10. *Botanical Safety Handbook* (American Product Association), Michel Mac Guffin *et al.*, CRC, Boca Raton, 1997.

11. *Toxicology of Botanical Medicines*, Francis Brinker, Eclectic Medical Publications, Sandy, Oregon, 2000.

12. University of Michigan Health System: http://www.umich/edu, 2010.

13. *Drug Herb Vitamin Interactions, Bible*, Richard Harkness and Steven Bratman, Prima Publishing, 2000.

14. *Handbook of Drug-Herb and Drug-Supplement Interactions*, Richard Harkness and Steven Bratman, Mosby, St Lois, Missourie, 2003.

15. *Adverse Effects of Herbal Drugs*-1(1991), 2(1993), 3(1995), Eds PAGM, De Smet *et al.*, Springer-Verlag Germany.

16a. *WHO Monographs on Selected Medicinal Plants*, Vol 1, 1999

16b. *WHO Monographs on Selected Medicinal Plants*, Vol 2, 2002

16c. *WHO Monographs on Selected Medicinal Plants*, Vol 3, 2007

16d. *WHO Monographs on Selected Medicinal Plants*, Vol 4, 2009

17.   British Herbal Compendium, Vol. 1, 1992; Vol. 2, 2006; Peter Bradley, British Herbal Medicine Association, Bournemouth.

PS: Relevant information from *"Safety Reviews on Selected Indian Medicinal Plants"* (Eds. Satyapal Singh Yadav and Neeraj Tandon, published by Indian Council of Medical Research, New Delhi) could not be incorporated in the text, as the project was already in final stages.

Detoxification Process of Toxic Ayurvedic Plant Drugs

Editor's Note

Classical procedures adopted by Ayurvedic *acharyas* need total revamping, if safe and effective Ayurvedic Plant Drugs are to be manufactured on a large scale commercially. Central Drug Research Institute should work out detoxifying procedures which can be followed by herbal pharmaceutical industry. After the main document, we have tried to suggest scientific detoxification process for a few toxic herbs of Ayurveda for further research.

Bhallataka Fruit
Semecarpus anacardium Linn.

Classical procedure: Soak in cow's urine for twenty four hours, wash thoroughly, thereafter soak in cow's milk, and wash with water. Repeat the process with cow's milk three times. Cut stalk portion and ground well in sifted brick powder. Wash in water. (CCRAS.)

Cut the nuts into 4 pieces without coming into contact with the seed oil. Soak for 3 days in cow dung solution. Wash with cold water, then with coconut water, and dry. (IMPCOPS, *Sahasrayoga* process.)

Adopted process: The fruits are soaked in Cow's urine, and Cow's milk. Thereafter, rubbed on brick gravels. After removing the thalamus portions, the fruits are kept either in Cow's urine (for 7 days) or Cow's milk (for 7 days), and are finally washed with water. The seeds are then sifted to a bag containing brick gravels (for 3 days), rubbed thoroughly and dried. During the process coconut oil is applied on the exposed body parts of the persons involved in the processing to reduce the chances of dermatitis.

Claimed outcome: Increase level of anacardol is observed in detoxified fruits in comparison to the raw fruits. A few studies prove the changes of *Rf* values of phytoconstituents in detoxified samples as compared to raw ones. Due to the decarboxylation of the oil, the anacardic acid gets converted into the less toxic anacardol. Detoxification does not affect the amount of total flavonoids and the total carbohydrate content; however, considerable decrease in total phenolic content was reported after the *shodhana* process. (Gajjar U *et al.,* Effect of *shodhana* process on quantity of phytoconstituents of *Semecarpus anacardium* Linn., *Int J Pharm Life Sci,* 2011, 2: 805-807.)

Antioxidant activity of *S. anacardium* decreases but the safety profile of the drug increases as the toxic phenolic oil is removed during *shodhana*. (Gajjar U *et al.,* Improvement in safety profile of *Semecarpus anacardium* Linn. by *Shodhana, Planta Med.* 2013,79: 64.)

Dhattura Seeds
Datura metel Linn.

Classical procedure: Digest mature seeds in cow's milk for 3 hours in *dola yantra* over mild fire. Wash with warm water and dry.

Adopted process: The seeds are soaked in freshly collected cow's urine and kept aside for 12 h. After washing, the seeds are transferred to the *dola yantra* for *svedana* process for 3 h. The seeds are again washed with lukewarm water, allowed to dry and the seeds testa are removed. (CCRAS.)

Soak the seeds in cow's urine for twelve hours and wash. Then gently pound to remove the husky exterior, dry and use. (IMPCOPS, *Ayurvedaprakasha* process.)

Claimed outcome: Reduction in total alkaloid content and increase in total protein content of the seed were observed after detoxification. Complete removal of scopolamine and partial removal of hyosciamine is reported.

Guggulu Gum-Resin
Commiphora mukul Hook. Ex. Stocks. Engl.

Classical procedure: Take course powder of 250 g of *Triphala* and 125 g of Guduchi (*Tinospora cordifolia*), add 4 litre of water, keep overnight. Next morning prepare a *kvaatha* by reducing to half. Boil the filtrate and dip in it 250 g of Guggulu in a muslin bag, suspended by a rod across the pan. During the boiling drench the bag with *kvaatha*, and stir constantly by a ladle. In about 10 to 12 operations the entire Guggulu will go into solution in the *kvaatha* when bag is empty of Guggulu, remove the bag and discard. Decant supernatant liquid leaving debris, if any. Concentrate the collected liquid and dry in shade. (CCRAS.)

Dissolve the drug in the decoction of the *Triphala* by heating, Filter and discard the insoluble matter. Heat until semisolid. Place on a mat in small bits and dry in sun. (IMPCOPS, *Dhanvantari nighantu* process.)

Adopted process: Guggulu is purified by *svedana* in *dola yantra* by using various media such as distilled water, *Triphala kvaatha,* cow's milk and cow's urine. When all the *Guggula* dissolves in media, muslin bag is to be removed and the liquid is evaporated to collect purified *Guggulu*.

Outcome: It is indicated in the literature that *shodhana* of *guggulu* may enhance specific action such as increasing mobile property, body tonic property, and bioavailability. *shodhita Guggulu* shows considerable antispasmodic activity against spasms induced by acetylcholine, histamine and barium chloride on ileum of guinea pig and Wistar rats, which are absent in *ashodhita guggula*. (Kamble R *et al.*, Evaluation of antispasmodic activity of different *shodhit guggulu* using different *shodhan* process, *Indian J Pharm Sci,*

2008, 70: 368-372.) The study demonstrates that purification processes significantly modified the anti-inflammatory activities of *Guggula*. (Karan M *et al.*, Effect of traditional Ayurvedic purification processes of guggulu on carrageenan-induced paw oedema in rats, *J Pharm Biomed Sci*, 2012, 21: 1-5.)

Gunjaa Seed
Abrus precatorius Linn.

Classical procedure: Seeds are put in *Kanjika* and *svedana* is done for 3 hours in *dola yantra*. The testa is removed. Thereafter the drug is washed and dries. (Ayurvedic Pharmacopoeia of India, Part I, Vol. 1: 200.)

The adopted process: The seeds are subjected to the *svedana* in *dola yantra* with cow's milk or *Kanji* for 3 to 6 h. The processed seeds are washed with hot water and then dried under shade.

Claimed outcome: High performance liquid chromatography (HPLC) study of the *Gunjaa* extract before and after the *shodhana* process showed that the level of toxic hypaphorine decreases, whereas the less toxic alkaloid abrine increases. Perhaps during *shodhana* process, a major part of hypaphorine might have Tundergone transformation into abrine by reduction of its tertiary amino group into the primary amino group. Percentage of protein present in *Gunjaa* also reduces after *shodhana*. (Singh G D *et al.*, Effect of *shodhana* on the toxicity of *Abrus precatorius*, *Anc Sci Life*, 1998,18: 127-129.)

Hingu latex
Ferula asafoetida Regel

Boil *ghrita* in a sauce pan, add clean dried latex of Hingu till fried to a brown colour. (CCRAS.)

Grind the drug in lime juice until glistening property is lost. (IMPCOPS.)

Jayapaal Seeds
Croton tiglium Linn.

Classical procedure: Reduce the drug to small pieces and soak in cow's urine in a closed vessel for 7 days, change cow's urine every twenty four hours for 7 days. Remove the skin of the seeds as well as cotyledonous coat, prepare shreds and dry. (CCRAS.)

Break open the seeds and collect the cotyledons discarding the embryonic part. Suspend the cotyledons in a cloth bag into a solution of cow dung and boil slowly for 4 hours, Allow it to cool and collect the cotyledons. Wash and dry them in sun. Then subject them to similar treatment using solution of cane jiggery and cow's milk. Finally wash. Dry and fry in *ghee*. (IMPCOPS, *Basavarajiya* process.)

Adopted process: Seeds are detoxified by *svedana* with cow's milk in a *dola yantra* for 3 h, after removing its raphae, triturated with lemon juice.

Outcome: The quantity of major purgative principles phorbol ester and crotonic acid in unpurified and purified samples were determined by HPLC. The content of the phorbol ester in unpurified and purified sample was found to be 5.2 mg/100 g and 1.8 mg/100 g of dried seeds of *C. tiglium,* respectively. The quantity of crotonic acid in unpurified seeds of *C. tiglium* was found to be 0.102 mg/100 g of dried seeds. Crotonic acid content was found to be absent in the purified seed extract of *C. tiglium.* (Pal PK *et al.,* Detoxification of *Croton tiglium* L. seeds by Ayurvedic process of *shodhana, Anc Sci Life,* 2014, 33: 157-161.)

Kararvira Leaf, Root
Nerium indicum Mill.

Classical procedure: The leaves or Roots are purified by *svedana* process in *dola yantra* using cow's milk for 3 h. Afterwards, the drugs are washed with water and dried.

Claimed outcome: There is a decrease in the cardenolide and oleandrin content. It is claimed that the detoxified herb showed no reported toxicity in animal models. (Banerjee A A *et al.,* Detoxification of *Nerium indicum* roots based on Indian system of medicine: Phytochemical and toxicity evaluations. *Acta Pol Pharm,* 2011, 68: 905-911.)

Laangali
Gloriosa superba Linn.

Classical procedure: Reduce the drug to small pieces, soak in cow's urine for twenty four hours, remove from cow's urine and dry in shade. (CCRAS, IMPCOPS.)

Outcome: After the *shodhana* process the level of colchicine is significantly reduce as colchicine is polar in nature and therefore soluble in *Gomutra* and water. (Nabar MP *et al., Gloriosa superba* roots: Content change of colchicine during *shodhana* (detoxification) process, *Indian J Tradit Knowl,* 2013, 12: 277-280.)

Snuhi Dugdha
Euphorbia neriifolia Linn.

Add 12 ml juice of Chinchaa (*Tamarandus indica* Linn.) fruit to 50 ml Snuhi dugdha, mix well and filter through cloth. Dry the liquid portion in sun and use the sediment. (CCRAS.)

Vachaa
Acorus calamus Linn.

Though *Vachaa* does not feature among 14 poisonous substances of Ayurveda in India's Drugs and Cosmetic Act, yet *Ayurvedic Pharmacopoeia of India,* Part I, Vol. II, page 170, recommended *shodhana* process for *Vachaa* rhizome before internal use.

Classical procedure: Authentic text could not be traced.

Adopted process: Vachaa is boiled successively in cow's urine, Mundi *kvaatha* (prepared from the whole plant of *Sphaeranthus indicus* Linn.) and *Pancha-pallava kvaatha* for 3 h. After this, the rhizome is treated with *Gandhodaka* for 1 h and the detoxified rhizomes are shade dried for 12 days.

Outcome: Toxicity studies indicate that oral administration of rhizomes of both *shodhita* and raw *Vachaa* powder at 2000 mg/kg in albino rats is relatively safe. (Bhat DS *et al.*, A comparative acute toxicity evaluation of raw and classically processed rhizomes of Vacha (*Acorus calamus* Linn.), *Indian J Nat Prod Resour*, 2012, 3: 506-511.)

Vatsanaabha Root
Aconitum ferox Wall.

Classical procedure: Reduce the Aconite (*Aconitum ferox* Wall., *Aconitum napellus* Linn., and *Aconitum chasmanthum* Holmes ex Stapf) root to small pieces and soak in cow's urine for 3 days in a closed vessel, changing the cow's urine after every twenty four hours. Remove cow's urine, wash the drug with water, digest with cow's milk in *dola yantra* for 3 hours, and dry thereafter. (CCRAS.)

Soak the drug pieces in cow's urine for 3 days, changing the urine every day. Dry them in the sun. Put in a cloth bag, suspend in cow's urine and boil for six hours. Wash the pieces in water and dry. (IMPCOPS, *Sahasrayoga* process.)

The *adopted process:* The root is subjected to *svedana* (boiling) in *dola yantra* using milk for 3 h daily for three continuous days, followed by washing with water thrice and drying under sun light.

Claimed outcome: After this process, the total alkaloid content decreases. (Sarkar PK *et al.*, *Indian J Pharm Educ Res*, 2012, 46: 243–247.) The contents of less toxic substances such as aconine, hypoaconine, and benzylhypoaconine increases. (Deore SL *et al.*, *J Young Pharm*, 2013, 5: 3-6), *possibly* due to conversion of the toxic aconitine into aconine or hydrolysis of the alkaloids to their respective amino alcohols. It has been claimed that cow's urine converts *Aconite* to a compound with cardiac stimulant property, whereas, raw *Aconite* showed cardiac depressant properties. The unpurified aconite root group showed significant increase in heart rate, increase in QRS complex time and increase in QT interval, however these parameters were statistically insignificant in purified aconite root treated group.(Paul A, Effects of Avurvedic *shodha*(processing) on dried tuberous Aconite, *Aconitum napellus,* root, *Indones J Pharm*, 2013, 24: 40–46.)

Vijayaa Leaves
Cannabis sativa Linn.

Classical procedure: Put Vijayaa leaves in a muslin bag and wash in water repeatedly till green colouration to the washings ceases. Thereafter, dry the material in shade within muslin bag in thin layer. (CCRAS.)

Adopted process: The leaves are boiled with *Babbula tvak kvaatha* for 3 h and the powder obtained is triturated with cow's milk. Toxic effects of can also be reduced by triturating with *Babbula tvak kvaatha* and frying the powder obtained in cow *ghee.* (Shastri K, *Rasa Tarangini*, pp. 651-652.)

Vishamushti Seed

Strychnos nux-vomica Linn.

Classical procedure: Reduce the drug to small pieces and soak in cow's urine in a closed vessel for 7 days, change cow's urine every twenty four hours for 7 days. Remove the skin of the seeds as well as cotyledonous coat, prepare shreds and dry.

Cut the seed into 4 pieces. Boil the decoction of *Amaranthus* roots, Dry in sun and finally fry in *ghee*. (IMPCOPS, *Basavarajiya* process.)

Adopted process: The purification of seeds includes soaking in liquid media (one after another) for 3–20 days. The liquid media include *kanji* (soaking for 3 days), cow's milk (boiling for 3 h), cow's urine (7 days soaking) and cow's *ghee* (fried till brownish red in colour and swollen.) Some traditional practitioners use castor oil instead of *ghee* for frying or immerse the seeds in the exudates scraped from the fresh leaves and stems of *Aloe vera* for 15 days, followed by ginger juice for 7 days. After detoxification process, the seeds are washed with lukewarm water where the outer seed coat and embryo are removed from the cotyledons.

Claimed outcome: The preliminary phytochemical investigation showed changes in the level of phytoconstituents in different methods of detoxification. Being acidic in nature, *kanji* is considered a better extraction medium because it may facilitate the extraction of alkaloids and other phytochemicals. Ginger juice also showed reduction in alkaloids present in the seeds. Cow's urine shows better pharmacological potency than the raw seeds.

The detoxification study of *S. nux-vomica* seeds was performed by Dr Chandra Kant Katiyar *et al.*, by traditional methods using aloe and ginger juice, by frying in cow ghee and by boiling in cow milk. All the treated samples were extracted with ethanol. Ethanol extracts were used for the evaluation of spontaneous motor acting (SMA), pentobarbitone-induced hypnosis, pentylenetetrazole (PTZ)-induced convoulsions, diazepam-assisted protection and morphine induced catalepsy. Ethanolic extracts of all the samples reduced SMA and inhibited catalepsy, but seeds processed in milk showed the lowest content of strychnine, exhibited marked inhibition of PTZ induced convulsion and maximum potentiation of hypnosis. (*Fitoterapia,* 2010, 81: 190-195.)

Ayurvedic Sources

Central Council for Research in Ayurvedic Sciences (CCRAS), *The Ayurvedic Pharmacopoeia of India,* Part I, Vol. I, *Vaidya Yoga Ratnavali,* The Indian Medical Practioner's Co-operative Pharmacy and Stores Ltd. Chennai, Tamil Nadu (IMPCOPS.), and Review paper by Santosh Kumar Maurya *et al., Shodhana:* An Ayurvedic process for detoxification and modification of therapeutic activities of poisonous medicinal plants, *Anc Sci Life,* 2015 Apr-Jun,34(4): 188-197.)

Suggestions for Detoxifying Toxic Ayurvedic Plant Drugs by Modern Methods

Suggestions for Detoxifying Toxic Ayurvedic Plant Drugs by Modern Methods

If extracts of toxic herbs are to be produced in a modern setting under a Traditional Regulatory System such as AYUSH, then the classical detoxification process is very cumbersome, and modern procedures are to be adopted.

The detoxification of the seeds of *Strychnos nux-vomica* the modern method involves boiling the seeds in water at high temperature, adding 20 per cent Beeswax to the mass and boiling for 8 hours. The mass is left to cool for 12 to 16 hours. The beeswax forms a hard layer on top of the aqueous phase, This wax layer is then removed and the water is tested for traces of strychnine, if there is a positive result the process is repeated a second time.

In the case of *Aconitum* spp., the detoxification process involves charcoal purification or resin purification. The solution of the extract is passed through activated charcoal or a resin column specific to the nature of alkaloid desired to be absorbed. An alternative method involves detoxifying with pyrolytic and hydrolytic pre-treatments, which requires repeated soaking in salt water, boiling until the roots turn black, and drying in the oven.

Anacardic acid from Cashew *Anacardium occidentale* nut shell liquid is treated with a basic lead compound whereupon the anacardic acid precipitates as crude lead anacardate. This precipitate is separated by any suitable means such as filtering, centrifuging, *etc.* and following separation, it is converted into crude anacardic acid by treating with acid. In the other method, cashew nut shell liquid is treated with a basic sodium, or other alkali, compound whereupon a gel composed of sodium anacardate and the nonacidic portion of the liquid is formed. The gel is leached with a suitable solvent to dissolve and remove the greater proportion of the non-acidic material from the sodium anacardate, and, following leaching, the crude sodium anacardate is converted into crude anacardic acid by treatment with acid.

These examples indicate that there is a possibility of detoxifying certain toxic compounds from medicinally valuable herbs by processing them under strictly controlled conditions in pharmaceutical companies for mass production of safe and Ayurvedic drugs.

The Editor is committed not to disclose the source.

Appendices

Appendix I

Ayurvedic Pharmacopoeia of India Plants in European Food Safety Authority (EFSA) List. (Substances of possible concern for human health when used in food and food supplements.)

Sl. No.	Ayurvedic Names	Botanical names
1.	Ashwagandha	Withania somnifera Dunal
2.	Atasi	Linun usitatssimum Linn.
3.	Ativisha	Aconitum heterophyllum Wall.
4.	Bilva	Aegle marmelos Corr. ex Roxb.
5.	Erand	Ricinus communis Linn.
6.	Gokshura	Tribulus terrestris Linn.
7.	Guduchi	Tinospora cordifolia (Wild) Miers
8.	Guggulu	Commiphora wightii (Arn.) Bhand
9.	Gunja	Arbus precatorius Linn.
10.	Hingu	Ferual foetida Regel
11.	Jatiphala	Myristica fragran Houtt.
12.	Kantakari	Solanum surattense Burm.f.
13.	Karavira	Nerium indicum Mill.
14.	Kushtha	Saussurea lappa C.B. Clarke
15.	Kutaja	Holarrhena antidysenterica (L.) Wall.
16.	Lavanga	Syzygium aromatium (Linn.) Merr. & Perry
17.	Misreya	Foeniculum vulgare Mill.
18.	Shunthi	Zingiber officinale Roxb.
19.	Tvak	Cinnamomum zeylanicum Breyn.
20.	Upakunchika	Nigella sativa Linn.
21.	Vidanga	Embelia ribes Burm.f.
22.	Vijaya	Cannabis sativa Linn.
23.	Yashtimadhu	Glycyrrhiza glabra Linn.
24.	Yavani	Trachyspermum ammi (Linn.) Sprague

Sl. No.	*Ayurvedic Names*	*Botanical names*
25.	Akarkarabha	*Anacyclus pyrethrum* DC.
26.	Bhallataka	*Somecarus anacardium* Linn.
27.	Gambhari	*Gmelina arborea* Roxb.
28.	Jayapala	*Croton tiglium* Linn.
29.	Karavellaka	*Momordica charantia* Linn.
30.	Nimba	*Azadirachta indica* A. Juss
31.	Vatsanabh	*Aconitum chasmanthum* Stapf ex Holmes
32.	Amra	*Mangifera indica* Linn.
33.	Arka	*Calotropis procera* (Ait.) R.Br.
34.	Atmagupta	*Mucuna prurita* Hook.
35.	Dhattura	*Datura metel* Linn.
36.	Isvari	*Aristolochia indica* Linn.
37.	Koshataki	*Luffa acutangula* (Linn.) Roxb.
38.	Gandhapuro	*Gaultheria fragrantissima* Wall.
39.	Mentha (Mint)	*Mentha* spp.
40.	Manjishtha	*Rubia cordifolia* Linn.
41.	Sthauneyak	*Taxus baccata* Linn.
42.	Kataka	*Strychnos potatorum* Linn.f
43.	Mahanimba	*Melia azedarach* Linn.
44.	Vasa	*Adhatoda zeylanica* Medic.
45.	Vishamushti	*Strychnos nux-vomica* Linn.
46.	Dravanti	*Jatropha glandulifera* Roxb.
47.	Elavaluka	*Prunus avium* Linn.
48.	Ahiphena	*Papaver somniferum* Linn.
49.	Kulanjana	*Alpinia galanga* Willd.
50.	Sarpagandha	*Rauvolfia serpentine* (Linn.) Benth. ex Kurz
51.	Vasa	*Adhatoda vasica* Nees.
52.	Daruharidra	*Berberis aristata* DC.

Due to repeated noncompliance, spices and herbs from certain countries (especially Indonesia and India) have been subject to additional and stricter custom controls in the European Union over the past years. Measures imposed by the European Union include requiring a health certificate and an analytical test report. (Courtesy: Dr C.K. Katiyar.)

Banned and Restricted Indian Herbal Ingredients in U.K.

Herbs which can only be sold in premises which are registered pharmacies and by or under the supervision of a pharmacist:

Aconitum napellus, Aconitum chasmanthum, Aconitum spicatum

Areca catechu

Aristolochia, Aristolochia clematis, Aristolochia contorta, Aristolochia debelis, Aristolochia fang-chi, Aristolochia manshuriensis, Aristolochia serpentaria (Indian species not mentioned.)

Artemisia cina (Indian species not mentioned.)

Atropa belladonna, Atropa acuminata herb

Atropa belladonna, Atropa acuminata root

Chenopodium ambrosioides var. *anthelminticum*

Cinchona officinalis

Clematis armandii, Clematis montana (Indian sp. *Clematis gouriana* not mentioned.)

Claviceps purpurea (Ergot of Rye)

Colchicum autumnale

Conium maculatum leaf and fruits

Convallaria majalis

Crotalaria berberoana, Crotalaria fulva, Crotalaria spectabilis Crotalaria spect (Indian sp. *Crotalaria juncea* or *Crotalaria verrucosa* not mentioned.)

Curcurbita maxima

Datura stramonium, Datura innoxia

Delphinium staphisagria

Digitalis leaf, *Digitalis* prepared

Dryopteris filix-mas

Embelia ribes, Embelia robusta

Ephedra gerardiana

Holarrhena antidysenterica

Hyoscyamus niger

Juniperus sabina

Lobelia inflata

Mallotus philippinensis

Mandragora autumnalis

Papaver somniterum

Pilocarpus jaborandi, Pilocarpus microphyllus

Podophyllum hexandrum

Punica granatum bark

Rauvolfia serpentina

Senecio vulgaris

Strophanthus kombe

Strychnos ignatii bean

Strychnos nux-vomica seed

Ulmus fulva

Veratrum viride

Viscum album

Source

https //www.gov.uk › List of banned or restricted herbal ingredients for medicinal use, Dec 18, 2014.

Appendix III

International Trade of Herbal Drugs Regulatory Overview
Dr Sanjay Sharma

International Trade of Herbal Drugs is subject to compliance with the International treaties like Convention on International Trade in Endangered Species of Wild Fauna and Flora (CITES.)[1] CITES, regulate the international trade of certain species, which are threatened with extinction or which would be reaching the status of being endangered, if their overexploitation continues.

Internationally, herbal products are regulated under different classifications, some of which are:

- Complimentary medicines
- Natural health products
- Prescription medicines
- Over the counter medicines
- Supplements
- Traditional herbal medicines.

The regulatory requirements of these vary considerably. While prescription medicines are strictly regulated, the extent of control on supplements is relatively low.

A review of the regulatory status of the herbal medicines across the globe is presented in this paper.

India[2]

Herbal drugs are regulated under the Drug and Cosmetic Act (D and C) 1940 and Rules 1945 in India, where regulatory provisions for Ayurveda, Unani, Siddha medicine are clearly laid down. Department of AYUSH is the regulatory authority and mandate that any manufacture or marketing of herbal drugs is to be done after obtaining manufacturing license, as applicable.

The D and C Act extends the control over licensing, formulation composition, manufacture, labeling, packing, quality, and export. Schedule "T" of the act lays down the good manufacturing practice (GMP) requirements to be followed for the manufacture of herbal medicines. The official pharmacopoeias and formularies are available for the quality standards of the medicines. First schedule of the D and C Act has listed authorized texts, which have to be followed for licensing any herbal product under the two categories:

- ASU (Ayurvedic, Siddha, Unani drugs)
- Patent or proprietary medicines.

Malaysia[3]

Herbal products in Malaysia fall under the category of regulated products. Any marketer intending to place the herbal products in the market require to register the product first. The applicant is required to be registered with the Malaysia Registrar of Business or Suruhanjaya Syarikat Malaysia under two classifications:

- Traditional products
- Health supplements.

While the authorities mandate only labeling "traditionally used for" in front of any claim made on the traditional product, only those functional claims, which are listed by the authority are allowed in supplements.

Philippines[4]

The herbal medicines are regulated in the Philippines as traditionally used herbal products. The regulators require that the preparations from plant materials, whose claimed application is based only on traditional experience of long usage, which should be at least five or more decades as documented in medical, historical, and ethnological literature are permitted to be marketed under this category.

The Bureau of Food and Drugs (BFAD), the regulatory authority in the country, mandate registration of the traditionally used herbal products before manufacture, import or market. The extent of control of BFAD includes the brand names of the traditional herbal products as well, and their prior clearance is required, before filing for product registration.

Authentication of the plant specimen needs to be obtained from the Philippine National Museum or any BFAD recognized taxonomist, and for imported products, the certificate of authenticity of the plants from the authorized government agency of the country of origin is accepted. The quality control requirements further lay down the pharmacopoeial standards. BFAD further mandates that product indications should not require supervision by a physician.

Nigeria

In Nigeria, the trade of herbal products is regulated by National Agency for Food and Drug Administration and Control (NAFDAC) who has classified these products as "Herbal Medicines and Related Products." Pre-marketing registration of herbal medicines and related products is mandatory in Nigeria.[5] All advertisements require a pre-clearance from NAFDAC.[6] No advertisement can be made claiming cure for any disease conditions listed in "Schedule 1" of the Food and Drug Act 1990.[7]

Saudi Arabia[8]

Herbal products are classified in Saudi Arabia as traditional products. They are allowed if they have at least 50 consecutive years of traditional use. Their dose and the method of preparation must be same as those used, traditionally.

According to the evidence provided, they may fall under the sub-categories:

- Pharmacopoeial evidence for traditional products
- Nonpharmacopoeial evidence for traditional products.

For the former, the medicinal ingredients, quantity, recommended dose, route of administration, duration of use, dosage form, directions of use, risk information should be same as in the Pharmacopoeia and the method of preparation must be traditional.

For the latter category, any two independent references must be provided to supplement the evidence supporting the safety and efficacy of the product, from clinical studies, pharmacopoeias, and textbooks references, peer-reviewed published articles, data from nonclinical studies on pharmacokinetics, pharmacodynamics, toxicity information, reproductive effects, and the potential genotoxicity or carcinogenicity of an ingredient or information based on previous marketing experience of a finished product.

Australia[9]

Therapeutic Goods Administration, the regulatory agency of Australia, regulates herbal products under the category of complementary medicine. Ayurvedic medicine, traditional Chinese medicine, and Australian indigenous medicines, all are covered under this category.

Complementary medicines which do not require medical supervision are permitted and have to be entered on the Australian Register for Therapeutic Goods (ARTG) before marketing. The low-risk medicines require to be listed while the medicines for comparatively higher risk therapeutic conditions require registration on the ARTG. Only evidence-based claims which are entered on the ARTG are allowed.

United States of America[10,11]

The botanical products are classified as a drug, food or a dietary supplement by the United States Food and Drug Administration on the basis of the claims or end use. A product that is used to prevent, diagnose, mitigate, treat or cure a disease would fall under the category of drug. If the intended use of a botanical product is to affect the structure or function of the human body, it may be classified as either a drug or a dietary supplement. As per FDA, the drug must be marketed under an approved New Drug Application (NDA.)

FDA regulates the dietary supplements under the Dietary Supplement Health and Education Act of 1994. These do not require premarket approval and it is the responsibility of the marketer to ensure the safety and labeling compliance of their products with the regulations. The claims need to comply with the regulatory guidelines issued by the FDA. The manufacturing of dietary supplements should be done as per the current GMP for dietary supplements.

Canada[12]

Since January 1, 2004, Health Canada regulates herbal remedies and traditional medicines such as Ayurvedic medicine, under the natural health products regulations. The regulations mandate that a manufacturer, packer, labeler or importer need to have a prior registration with Health Canada before commencing any such activity.

The process involves registration of the manufacturing site/s along with the products. Complete data on product composition, standardization, stability, microbial and chemical contaminant testing methods and tolerance limits, safety and efficacy along with ingredient characterization, quantification by assay or by input needs to be submitted to Natural Health Product Directorate (NHPD.) The authority mandate that NHPs must comply with the contaminant limits and must be manufactured as per the GMP norms.

European Union[13]

The European Medicine Agency have laid down two ways of registration of herbal medicinal products: (1) A full marketing authorization by submission of a dossier, which provides the information on quality, safety and efficacy of the medicinal products including the physico-chemical, biological or microbial tests and pharmacological, toxicological and clinical trials data; under directive 2001/83/EC. (2) For traditional herbal medicinal products, which do not require medical supervision, and where evidence of long traditional use of medicinal products exists, and adequate scientific literature to demonstrate a well-established medicinal use cannot be provided, a simplified procedure under directive 2004/24/EC exists.

The evidence of traditional use is accepted as evidence of efficacy of the product. However, authorities may still ask for evidence to support safety. Quality control requirements require physico-chemical and microbiological tests to be included in the product specifications. The product should comply to the quality standards in relevant pharmacopoeias of the member state or European Pharmacopoeia. The bibliographic evidence should support that the product has been in medicinal use for at least 30 years, including at least 15 years within the European community. The application for traditional use registration shall be referred to the Committee for Herbal Medicinal Products, if the product has been in the community for less than 15 years, but otherwise qualifies for the simplified registration procedure under the directive.

Editor's Note: Dr Sanjay Sharma has specialized in International Regulatory Status of Herbal Drugs.

References

1. Geneva, Switzerland: 2014. [Last cited on 2014 Apr 29]. Convention on International Trade in Endangered Species of Wild Fauna and Flora. Available from: http: //www.cites.org/eng/disc/what.php.

2. Malik V, editor, 23rd ed. Lucknow: Eastern Book Company; 2013. Law Relating to Drugs and Cosmetics.

3. Ishak R, Mohamad J. Guidelines on Registration of Traditional and Health Supplement Products, Revised ed. Version 1.0. Kuala Lumpur. Malaysian Biotechnology Corporation SDN BH. 2011. [Last cited on 2014 Apr 29]. Available from: http: //www.biotechcorp.com.my/wp.content/uploads/2011/11/guidelines-on-registration-of-traditional-health-supplement-product-Dec-2011-Revision.pdf.

4. Geneva: WHO Library Cataloguing-in-Publication Data, World Health Organization; 2005. [Last cited on 2014 Apr 29]. Guidelines on the Registration of Traditionally used Herbal Products, Department of Health, Republic of Philippines, Report of a WHO Global Survey. Available from: http: //www.apps.who.int/medicinedocs/pdf/s7916e/s7916e.

5. Nigeria: Herbal Medicines and Related Products (Registration) Regulations; 2005. [Last cited on 2014 Apr 29]. National Agency for Food and Drug Administration and Control (NAFDAC) Available from: *http: //www.wipo.int/wipolex/en/text.jsp?file_id=218178.*

6. Nigeria: Herbal Medicines and Related Products (Advertisement) Regulations; 2005. [Last cited on 2014 Apr 29]. National Agency for Food and Drug Administration and Control (NAFDAC) Available from: http: //www.nafdac.gov.ng/attachments/article/205/14_HERBAL_MEDICINES_AND_RELATED_PRODUCTS_ADVERTISMENT_REGULATIONS_2004.

7. Nigeria: Herbal Medicines and Related Products, (Labeling) Regulations; 2005. [Last cited on 2014 Apr 29]. National Agency for Food and Drug Administration and Control (NAFDAC) Available from: http: //www.wipo.int/wipolex/en/text.jsp?file_id=218178.

8. Saudi Food and Drug Authority. Kingdom of Saudi Arabia: Data Requirements for Herbal and Health Products Submission: Contents of Dossier, Version 1. 2012. [Last cited on 2014 Apr 29]. Available from: http: //www.old.sfda.gov.sa/NR/rdonlyres/90675DEA-E1D0-4869-95E6-EED33281270D/0/DataRequirementsforHerbalandHealthproductssubmission_2012.pdf.

9. North Sydney: Therapeutic Goods Administration; 2014. [Last updated on 2014 Apr 09; Last cited on 2014 Apr 29]. Australian Government Department of Health. Available from: https: //www.tga.gov.au/

10. Food and Drug Administration. Rockville: U.S. Department of Health and Human Services, Center

for Drug Evaluation and Research (CDER): Guidance for Industry, Botanical Drug Products Online Resource. 2004. [Last cited on 2014 Apr 29]. Available from: http: //www.fda.gov/cder/guidance/index.htm.

11. Food and Drug Administration. Rockville: U.S. Department of Health and Human Services, Dietary Supplements Guidance Documents and Regulatory Information Online Resource. 2014. [Last cited on 2014 Apr 29]. Available from: http: //www.fda.gov/Food/GuidanceRegulation/GuidanceDocumentsRegulatoryInformation/DietarySupplements/default.htm.

12. Health Canada. Ottawa: Natural Health Products–Drugs and Health Products. 2013. [Last cited on 2014 Apr 29]. Available from: http: //www.hc-sc.gc.ca/dhp-mps/prodnatur/index.eng.php.

13. Official Journal of the European Union: Directive 2004/24/EC; 31st March. 2004. [Last cited on 2014 Apr 29]. Available from: http: //www.eurlex.europa.eu/LexUriServ/LexUriServ.do?uri=OJ: L: 2004: 136: 0085: 0090: en: pdf.

Editor's Note: The references will help in updating the status of Herbal products in international trade.

Appendix IV

Abbreviations

α: alpha
β: beta
γ: gamma
ABTS: 2,2'-azino-bis(3-ethylbenzothiazoline-6-sulphonic acid)
ACE: Angiotensin converting enzyme
AFI: Ayurvedic Formulatory of India
API: Ayurvedic Pharmacopoeia of India
BHA: Butylated hydroxyanisole
BHT: Butylated hydroxytoluene
bw: Body weight
cm: Centimeter
CCl_4: Carbon tetrachloride
CCRAS: Central Council for Research in Ayurveda and Siddha
COX: Cyclooxygenase
CNS: Central Nervous System
CTC: Common toxicity criteria
CVS: Cardiovascular system
d: day(s)
DNA: Deoxyribonucleic acid
L-Dopa: Levodopa

E. coli: *Escherichia coli*

ED_{50}: Median effective dose

EDTA: Ethylenediaminetetraacetic acid

FSH: Follicle stimulating hormone

GABA: Gamma-amino butyric acid

g/gm: Gram(s)

g/kg: Gram per kilogram

h: Hour

Hb: Haemoglobin

HDL: High density lipoproteins

HIV: Human immunodeficiency virus

HPLC: High pressure liquid chromatography

HPTLC: High performance thin layer chromatography

HSV-1,-2: *Herpes simplex virus* 1 and 2

5-HT: 5-Hydroxytryptamine

i.m.: Intramuscular

i.p.: Intraperitoneal

i.v.: Intravenous

IC_{50}: Median inhibitory concentration

ICMR: Indian Council of Medical Research

IMPCOPS: The Indian Medical Practitioner's Co-operative Pharmacy and Stores Ltd.

IU: International Unit

ID_{50}: Median inhibitory dose

IP: Indian Pharmacopoeia

Kcal/kg: Kilocalorie per kilogram

LC_{50}: Median lethal concentration

LD_{50}: Median lethal dose

LDL: Low density lipoproteins

LH: *Luteinizing hormone*

m: Meter

MIC: Minimum inhibitory concentration

MTD: Maximum tolerated dose

µg: Microgram

mg: Milligram(s)

mg/kg: Milligram per kilogram

ml/mLvMillilitre

NLT: Not Less Than

NMT: Not More Than

p.o.: Per oral

PMID: PubMed identifier unique number.

ppm: Parts per million

RBC: Red blood corpuscles

s.c.: Subcutaneous

SGOT: Serum glutamic oxaloacetic transaminase

SGPT: Serum glutamic-pyruvic transaminase

Sh.: Shigella

Sp.: Species

Spp.: Multiple species

Subsp.: Subspecies

Staph.: *Staphylococcus*

Syn.: Synonym

TLC: Thin layer chromatography

UV: Ultraviolet

Var.: Variety

Vib.: *Vibrio*

VLDL: Very low density lipoproteins

v/v: Volume per volume

v/w: Volume per weight

WBC: White blood corpuscles

Wk: Week(s)

w/w: weight per weight

Index

Index

I

Illicium anisatum: The Adulterant of *I. verum* 92

Illiciun verum 92

Indian Aconite 160

Indian Asparagus 31, 166

Indian Barberry 37, 167

Indian Bdellium 60, 170

Indian Birthwort 165

Indian Cinnamon 56

Indian Dill 163

Indian Ginseng 185

Indian Gum Arabic Tree 159

Indian Hemp 45, 168

Indian Kudzu 123

Indian Long Pepper 178

Indian Madder 129

Indian Olibanum 167

Indian pennywort 55

Indian Rhubarb 180

Indian Sarsaparilla 174

Indian Snakeroot 126, 180

Indian species of Withania 153

Indian Squill 184

Indian Tobacco 104

Indian Valerian 184

Indigo 94

Indigofera tinctoria 94

Indravaaruni 59, 170

Ingudi 35

Inhibitory Effects of *Asparagus racemosus* on the Digestive Enzymes 31

Inotropic Effect of an Ethanol Extract of Guava Leaves 122

Interaction of Capsaicin with Drug-Metabolization 47

Interaction of *Cinnamomum cassia* with Tetracycline 58

Interaction of *Cinnamomum tamala* with Gentamicin 57

Interaction of Dhatura with Succinylcholine 73

Interaction of Garlic with Drugs 14

Interaction of Pomegranate Juice with Rosuvastatin 124

Interactions of Aloe with Sevoflurane 17

Interactions of Berberine with Drugs 38

Interactions of Menthol and Peppermint Oil with Drugs 106

Inula racemosa 175

Invalidated Claims of Herbal Antidotes for Poisoning 11

Irimeda 160

Ishvari 165

J

Jaatiphala 110, 177

Jackfruit 30, 165

Jambu 146, 182

Jatamansi 111, 127, 177

Jatropha curcas 95

Java Plum 146, 182

Jayanti 133

Jayapaal 66, 171

Jayapaal Seeds 193

Jeeraka Krishna 168

www.ingramcontent.com/pod-product-compliance
Lightning Source LLC
Chambersburg PA
CBHW031949180326
41458CB00006B/1667